Photonics of Quantum-Dot Nanomaterials and Devices

THEORY AND MODELLING

Photonics of Quantum-Dot Nanomaterials and Devices

THEORY AND MODELLING

ORTWIN HESS
Imperial College London, UK

EDELTRAUD GEHRIG
Hochschule Heilbronn, Germany

Imperial College Press

ICP

Published by

Imperial College Press
57 Shelton Street
Covent Garden
London WC2H 9HE

Distributed by

World Scientific Publishing Co. Pte. Ltd.
5 Toh Tuck Link, Singapore 596224
USA office: 27 Warren Street, Suite 401-402, Hackensack, NJ 07601
UK office: 57 Shelton Street, Covent Garden, London WC2H 9HE

British Library Cataloguing-in-Publication Data
A catalogue record for this book is available from the British Library.

PHOTONICS OF QUANTUM-DOT NANOMATERIALS AND DEVICES
Theory and Modelling

ISBN-13 978-1-84816-521-2
ISBN-10 1-84816-521-8

Typeset by Stallion Press
Email: enquiries@stallionpress.com

Printed in Singapore.

To Jens, Rebecca and Gabriel
And
To Heike, Rita, Francis, Killian and Maxwell

Contents

Chapter 1

Introduction to Photonic Quantum Dot Nanomaterials and Devices

Nanotechnology has revolutionised photonics, enabling us to conceive materials such as carbon nanotubes (CNT), quantum dots (QD) and metamaterials that have novel optical properties and enable functionalities right down at the materials level. This, however, has brought at least as many new challenges to our understanding in just how light interacts with matter on the nanoscale and also how matter is different when it comes in such small, nano- (but not atomic-) sized pieces. Indeed, for applications in optoelectronics, quantum dot nanomaterials show very promising characteristics allowing for the design of the spatial and spectral properties of propagating light fields. Using these materials for lasers (by applying a suitable electric contact and carrier injection) leads to a particularly high differential gain and low laser threshold current (Asada *et al.*, 1986), low alpha factor (Willatzen *et al.*, 1994) and ultrafast modulation properties (Kapon, 1999a). In this chapter we summarise the physical properties of photonic quantum dot nanomaterials and introduce the various types and configurations of photonic quantum dot devices. The theory that forms the basis for our study of the physics of photonic quantum dot nanomaterials and devices will be introduced in Chapter 2.

1.1 Physical Properties of Quantum Dots

The specific epitaxial structure of quantum dot nano systems implies a three-dimensional carrier confinement which leads to a spatial carrier localisation and discrete energy levels. As a consequence, ensembles of QD are very attractive for the development of novel gain media with properties specifically designed for a particular application. QD gain media that combine the advantages of semiconductors and atoms allowing for the design of highly coherent, tuneable and compact laser sources. Furthermore, compared to the GaAs-based material technology of bulk and quantum well structures, they allow generation of laser emission at longer wavelengths. Needless to say, this is of particular importance for telecommunication applications. The specific design of the active media also leads to many additional promising properties such as a reduced temperature sensitivity, atomic-like gain characteristics with reduced phase-amplitude coupling (α-factor) leading to good spatial and spectral purity, low chirp and reduced reflection sensitivity (reducing the need for optical isolators).

Semiconductor quantum dot nanomaterials are made by epitaxial growth technology, typically based on the principle of self-organised Stranski–Krastanov QD growth using molecular beam epitaxy (MBE) for exploratory materials development and the metal organic chemical vapour deposition (MOCVD) method meeting industrial standards. These methods allow the development of quantum dot materials at $1.3\,\mu$m and near $1.55\,\mu$m. In combination with a suitable design of wave guides, cladding, electronic contact and barrier layers, as well as of the laser cavity, allow the fabrication of photonic devices suitable for the industrial mass-market of information technology. In experiments, the physical properties (e.g. electronic structure, inhomogeneous broadening decay times) of quantum dot systems can be measured and detected with various methods: among the most important configurations are pump–probe setups (Berg *et al.*, 2001) and photoluminescence measurements (Chen *et al.*, 1998).

1.2 Active Semiconductor Gain Media

Generally, the fundamental principle of an active semiconductor medium is the recombination of electron-hole pairs in a suitable layer of semiconductor medium embedded in a chip forming an optical cavity for the optical fields propagating within it. Inversion of this medium is then realised by optical or electronic pumping via an optical pump beam, or directly via electrical contacts by applying a current. Once the pump beam or current reaches a characteristic threshold spontaneous emission processes are exceeded by stimulated emission and lasing starts.

The gain of an active semiconductor medium is the key element for the generation of coherent light realised in a laser configuration consisting of active gain medium, cavity and an external pump process. To achieve gain in semiconductor media, a characteristic density of electron-hole pairs has to be appropriately spatially localised. This may be realised by applying a current at the junction of a pn-diode. Today, most semiconductor (diode) lasers are based on III-V semiconductors. The active region of a typical system is based on GaAs and $Ga_{1-x}In_xAs$ where the subscript x indicates the fraction of Ga atoms in GaAs that is replaced by In. Typical nano-structures based on $Ga_{1-x}In_xAs$ emit — depending on expitaxial growth, structuring and doping — in the spectral window of 1.3 to 1.55 μm that is relevant for telecommunication. Without loss of generality, the numerical results presented in this book will refer to this material system and wavelength range. Depending on the geometry and thickness of the active layer one differentiates between bulk (heterostructure, three-dimensional '3D') or quantum well (two-dimensional '2D'), quantum wire (one-dimensional '1D'), and quantum dot (zero-dimensional '0D') nanostructures. Thereby, the dimensionality marks the number of dimensions the charge carriers effectively 'see'.

The physical properties of active nanomedia can be described on the basis of $\mathbf{k} \cdot \mathbf{p}$-theory. This approach allows the simulation of electronic and optical properties and naturally includes details on the size, shape and composition of the quantum dots. Typical geometries of quantum dots are pyramids or lens-shaped dots. Thereby, the vertical aspect ratio may, as well as the composition, vary from system to system. Since some of these parameters are known, or can only be measured very inaccurately, it is also possible to vary the most crucial structural properties like size, shape, composition or composition-profile systematically in the modelling. This eventually provides all essential electronic and optical properties that can be integrated in a space-time model.

1.3 Quantum Dot Lasers

Many novel laser structures use the promising properties of quantum dots. In these systems, a quantum dot layer is embedded in vertical and lateral directions in a multi-layer structure. The active medium represented by the quantum dot ensemble allows the generation of stimulated emission required for the lasing process. Combined with a form feedback by an optical resonator (realised by the natural reflectivities of the material or by a specific grating structure) one thus obtains a lasing structure with promising spatial and spectral emission properties. The excellent output characteristics of these nanomaterials thereby is a direct result of the strong carrier localisation and the discrete level structure characterising a quantum dot ensemble. In the following we will introduce two typical resonator structures employed in quantum dot laser devices. For details we refer the reader to the detailed discussion in (Coldren and Corzine, 1995; Diehl, 2000; Kapon, 1999a; 1999b).

1.3.1 *Heterostructure lasers*

Semiconductor (double) heterostructure lasers consist of at least three semiconductor layers sandwiched on top of each other (Thompson, 1980; Yariv, 1989). The middle layer, the active zone, has a smaller band-gap than the two outer ones, the cladding layers. These layers form a pn-junction. The thin active region is usually un-doped or slightly doped p-type, while one of the adjacent layers is heavily doped p-type and the other one n-type. Application of a positive bias current leads to an injection of electrons from the n-type layer into the active layer and to a creation of holes in the p-side. The potential barriers resulting from the energy difference between the energy gaps of the different compound semiconductors thereby prevent the diffusion of the charge carriers out of the active region into the p- or n-type layer. Furthermore, the energy of the optical mode is confined to the central active layer: the different doping of the layers leads to a spatial index profile and thus to the formation of a dielectric waveguide in the vertical direction for the optical fields propagating in the structure. Many of today's mass-produced semiconductor lasers still rely on the principle of the double confinement (of carriers and radiation) realised in the double heterostructure.

1.3.2 *Active nanomaterials*

Generally, a large number of semiconductor materials can be used to manufacture quantum structures. In addition, one can combine different semiconductors with favourable properties in an alloy. This variation of input parameters is usually referred to as band-gap engineering. Band-gap engineering opens fascinating possibilities, in particular for fabrication of novel laser structures (Yariv, 1989) and the tayloring of emission wavelengths. The best controlled III-V quantum system is the GaAs/AlGaAs structure with emission in the red range (Yariv, 1989). Another attractive material showing significant advantages over the conventional long wavelength structures is the quaternary alloy GaInAsN, which was first proposed by (Kondow *et al.*, 1996). Good temperature characteristics and simpler fabrication makes GaInAsN VCSELs very attractive for applications in high-speed optical networks (Nakatsuka *et al.*, 1998; Nakahara *et al.*, 1998; Sato *et al.*, 1997).

Today advanced semiconductor growth techniques allow realisation of a semiconductor structure with a precision down to a single atomic layer. This has pushed forward the development of mesoscopically structured semiconductor systems. If an active semiconductor layer is sandwiched between semiconductor material of larger band-gap a quantum well is formed between the barriers and a confinement of the carrier dynamics in one spatial direction occurs. Typical layer thicknesses are a few nm, i.e. a limited number of atomic layers. The restriction on the movement of the carriers in a two-dimensional plane leads to quantisation effects and thus also affects the carrier energy as compared to a 'free' electron in a bulk medium: the system now is characterised by allowed energy bands whose energy positions are dependent on the height and width of the barrier. Details of the energetic structure can be calculated by means of fundamental quantum mechanics.

Due to their discrete, atom-like energy level structure, semiconductor quantum dots have unique electronic and optical properties and thus are ideal for novel laser devices based on mesoscopically structured materials. The good spatial and spectral purity of these systems is a direct consequence of the confinement of the electronic wave-functions in all three spatial dimensions leading to discrete energy levels and to a strong localisation of carriers. In recent years, rapid progress in the epitaxial fabrication of self-assembled III/V QDs has triggered tremendous efforts to use them as a gain medium in semiconductor lasers (Bimberg *et al.*, 1998). QD lasers show many advantageous properties, such as low and almost temperature-independent threshold current densities, high material gain (Arakawa and Sakaki, 1982) and low amplitude-phase coupling (α-factor) (Diehl, 2000). Today, QD lasers exhibit threshold current densities similar to good quantum well (QW) lasers but show a much higher beam quality (Asada *et al.*, 1986). Furthermore, the potential of QDs for being used in lasers with high-power outputs and extended wavelength ranges has attracted additional interest. In particular, the possibility of fabricating QD lasers emitting in the 1.3 μm wavelength region that are grown on inexpensive GaAs substrates and can be intetrated with existing III/V technology appears extremely interesting for telecommunication applications (e.g. signal processing, switching, wavelength division multiplexing) in the spectral window of minimum dispersion in glass fibres.

1.4 Laser Cavities

Concerning the geometry of the laser resonator, one generally differentiates between the families of edge-emitters (or in-plane lasers) and vertical-cavity lasers (or surface-emitters).

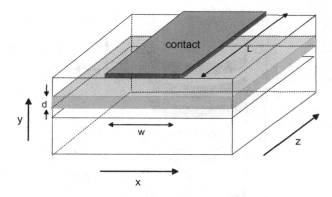

Fig. 1.1. Schematic of the geometry of an edge-emitting semiconductor laser. Charge carriers injected through the contact region at the top of the device (*dark grey*) recombine in the active zone. The active layers of the edge-emitter are located between the cladding layers. Light, generated by stimulated emission and amplification, propagates in the longitudinal (z) direction.

1.4.1 *In-plane edge-emitting lasers*

In an *edge-emitting* laser or amplifier (Fig. 1.1) carriers are injected via the contacts on the top. The carriers are captured by the dots embedded in the xy area. In the vertical direction, the dots are aligned in stacks (in approximately three to ten layers). Light generated by electron-hole recombination then travels in the longitudinal (z) direction inside the cavity formed by the two mirrors at the front and back. The length L of the cavity typically ranges from a few hundred μm to 3000 μm). This guarantees a sufficient gain during the propagation of the light fields in the resonator. The comparatively small thickness (about 0.1 μm) of the active layer in the vertical (y) direction resulting from the semiconductor epitaxial layer structure assures a vertical guiding of the optical waves. The transverse (x) may be considerably larger (about 3–5 μm for a single-mode laser and 50–200 μm in case of a multi-mode high-power broad-area laser). These typical dimensions of the active area strongly suppress one of the polarisation directions and thus lead to the emission of linear polarised light.

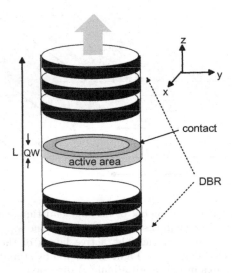

Fig. 1.2.   Schematic of the geometry of a vertical-cavity surface-emitting laser (VCSEL). The active layers of the VCSEL are located between distributed Bragg reflector (DBR) layers.

1.4.2   *Vertical-cavity surface-emitting lasers*

In *vertical-cavity surface-emitting* lasers (Fig. 1.2) (VCSELs), the geometry of the cavity is completely different. Most notably, the length of the resonator now only measures about one wavelength. Consequently, only a single longitudinal mode will be dynamically relevant and propagation effects may be disregarded. At the same time, both transverse (x and y) directions are equally large (typically 3–$30\,\mu$m). One has to therefore consider two transverse dimensions in theoretical approaches. Dielectric multilayers at both ends of the cavity lead to high mirror reflectivities and guarantee high gain. In the active area of the VCSEL, the recombination of an electron-hole pair leads, with equal probability, to the two possible polarisation directions. It is therefore, in particular, the resonator geometry or the epitaxial structure that determines the polarisation properties of the emitted radiation. Furthermore, due to the symmetric geometry, the polarisation of the emitted light is highly sensitive to the microscopic carrier and light field dynamics, anisotropies in the crystal structure or strain and optical anisotropies in the mirrors. As a consequence, VCSELs may exhibit polarisation instabilities in the input–output characteristics, which are the limiting factor in polarisation-sensitive applications. In theoretical approaches this can be considered by explicitly calculating the dynamics of microscopic dipoles and the light field dynamics for the two possible polarisations.

1.4.3 *High-power laser amplifiers*

Modern applications in nonlinear optics or telecommunications require laser sources offering good spatial and spectral purity, high output power and ultrafast response to high-speed modulation. The amplification of a coherent light single (e.g. a single-stripe laser) in the active area of an antireflection-coated large-area semiconductor laser (Bendelli *et al.*, 1991; Goldberg and Weller, 1991; Goldberg, *et al.*, 1992, 1993; Mehuys *et al.*, 1993, 1994; Mukai *et al.*, 1985; O'Brien *et al.*, 1997; Parke *et al.*, 1993; Sagawa *et al.*, 1996; Saitoh and Mukai, 1991) allows the generation of coherent light with such properties. During propagation, injected light basically maintains its spatial and spectral properties (Goldberg *et al.*, 1993). Up to now, various amplifier systems have been realised (broad-area amplifiers in single-pass or double-pass configuration, amplifiers with tapered geometry). In particular, the tapered amplifier (Fig. 1.3) has, due to its small signal gain and good wave-guiding properties, been the focus of theoretical (Hess and Kuhn, 1996; Lang *et al.*, 1993; Moloney *et al.*, 1997) and experimental (O'Brien *et al.*, 1997; Walpole, 1996) investigations. It consists of two parts, a single-stripe waveguide with a width of $\approx$3–5 μm and a tapered section in which the active area enlarges in the propagation direction so that the intensity at the output facet is kept below the threshold value for catastrophic optical mirror damage (COMD). The facets of the tapered amplifiers are anti-reflection coated. For good beam quality, this facet reflectivity should be less than 10^{-4} (Berg *et al.*, 2001). Alternatively, the wave propagation in the resonator should be off-axis, i.e. the facets should be angled with respect to the resonator axis. The small transverse dimension of the waveguide at the input facet of the active area leads

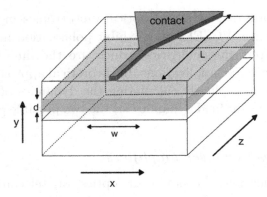

Fig. 1.3.   Scheme of a large-area semiconductor laser with tapered geometry.

to a high small-signal gain allowing efficient saturation of the inversion within the active layer for very moderate input powers of a few mW. Typical lengths of the small waveguide are a few $100\,\mu$m at a total length of 1–3 mm of the device.

1.4.4   *Coupled-cavity systems*

Coupled-cavity systems (Fig. 1.4) consist of two separately contacted sections in direction of light propagation, represent another promising configuration allowing for a separate optimisation of a pulse generation process (in the current-modulated short section) and the pulse amplification process (in the long second section).

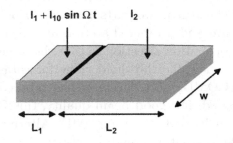

Fig. 1.4.   Scheme of a two-section laser.

1.4.5 *Optically excited nano systems*

The emission properties of semiconductor laser structures with large extension of the active area typically is strongly determined by the spatio-temporal coupling and interplay of light propagation, diffraction, carrier diffusion and microscopic carrier scattering processes leading to dynamic optical patterns (filamentation). Recently, the realisation of optically pumped semiconductor lasers (typically a vertical-cavity surface-emitting laser in external resonator configuration) has attracted attention. This concept allows for a combination of the power scaling involved in the high gain of semiconductor laser devices and the high beam quality provided by direct optical excitation. Optically pumped nano VCSELs thus combine the advantages of VCSELs and large-area laser amplifiers. They thus represent promising light sources for future laser technologies and applications. Furthermore, a spatially selective optical excitation and the resulting localisation or transfer of light may open the way to future quantum memories.

1.4.6 *QD metastructures*

In recent years, improvements in material technology have enabled the design of novel nano-structures with a specific spatial structuring combining the tailoring of the properties of propagating light fields with the design of the localised dot emitters. Embedding semiconductor quantum dots in micro-cavities allows an additional engineering of the photonic emission characteristics. In such a system the micro-cavity is of the order of the wavelength of the emitted light. As a consequence, compared to edge-emitting lasers the emission pattern and even the local spontaneous emission rate can be controlled more efficiently. Recent work on semiconductor quantum dot micro-cavities focused on the enhancement of spontaneous emission in microposts (Pelton *et al.*, 2002) or on an analysis of whispering-gallery modes in microdisks (Gayral *et al.*, 1999). Future systems on the basis of a semiconductor quantum dot in a micro-cavity small enough to contain only one mode resulting in a dot-in-a-dot system, may enable the generation of controlled single photon emission. This represents an important basis towards novel quantum communication and cryptography schemes (Waks *et al.*, 2002). Furthermore, micro-cavity lasers are also of great interest for future low power applications and, due to their fast response, to external pumping, potentially in high-speed optical communications.

References

Arakawa, Y. and Sakaki, H., *Appl. Phys. Lett.* **40**, 939–941, (1982).

Asada, M., Miyamoto, Y. and Suematsu, Y., *IEEE J. Quantum Elect.* **22**, 1915–1921, (1986).

Bendelli, G., Komori, K., Arai, S. and Suematsu, Y., *IEEE Photonic Tech. L.* **3**, 42–44, (1991).

Berg, T.W., Bischoff, S., Magnusdottir, I. and Mork, J., *IEEE Photonic Tech. L.* **13**, 541–543, (2001).

Bimberg, D., Grundmann, M. and Ledentsov, N.N., *Quantum Dot Heterostructures*. John Wiley, Chichester, (1998).

Chen, M.-C., Lin, H.-H. and Shie, C.-W., *J. Appl. Phys.* **83**, 3061–3064, (1998).

Coldren, L.A. and Corzine, S.W., *Diode Lasers and Photonic Integrated Circuits*. John Wiley, New York, (1995).

Diehl, R., (Ed.), *High-Power Diode Lasers: Fundamentals, Technology, Applications*. Springer, Berlin, (2000).

Gayral, B., Gerard, J.M., Lemaitre, A., Dupuis, C., Manin, L. and Pelouard, J.L., *Appl. Phys. Lett.* **75**, 1908–1910, (1999).

Gehrig, E., Hess, O. and Wallenstein, R., *IEEE J. Quantum Elect.*, **35**, 320–331, (1999).

Goldberg, L. and Weller, F., *Appl. Phys. Lett.* **58**, 1357–1359, (1991).

Goldberg, L., Mehuys, D. and Hall, D.C., *Electron. Lett.* **20**, 1082–1084, (1992).

Goldberg, L., Mehuys, D., Surette, M.R. and Hall, D.C., *IEEE J. Quantum. Elect.* **29**, 2028–2043, (1993).

Hess, O. and Kuhn, T., *Phys. Rev. A* **54**, 3360–3368, (1996).

Kapon, E. (Ed.), *Semiconductor Lasers I*. Academic Press, San Diego, (1999a).

Kapon, E. (Ed.), *Semiconductor Lasers II*. Academic Press, San Diego, (1999b).

Kondow, M., Uomi, K., Niwa, A., Kitatani, T., Watahiki, S. and Yazawa, Y., *Jpn. J. Appl. Phys.* **35**, 1273–1275, (1996).

Lang, R.J., Hardy, A., Parke, R., Mehuys, D., O'Brien, S., Major, J. and Welch, D., *IEEE J. Quantum Elect.* **29**, 2044–2051, (1993).

Mehuys, D., Goldberg, L. and Welch, D.F., *IEEE Photonic Tech. L.* **5**, 1179–1182, (1993).

Mehuys, D., Welch, D.F. and Goldberg, L., *Electron. Lett.* **28**, 1944–1946, (1994).

Moloney, J.V., Indik, R.A. and Ning, C.Z., *IEEE Photonic Tech. L.* **9**, 731–733, (1997).

Mukai, T., Yamamoto, Y., Kimura, T. and Tsang, W.T., *Semiconductors and Semimetals*, Vol. 22, chapter Opitcal Amplification by Semiconductor Lasers. Lighwave Communications Tech., Part E, Integrated Optoelectronics, (1985).

Nakahara, K., Kondow, M., Kitatani, T., Larson, M.C. and Uomi, K., *IEEE Photonic. Tech. L.* **10**, 487–488, (1998).

Nakatsuka, S., Kondow, M., Kitatani, T., Yazawa, Y. and Okai, M., *Jap. J. Appl. Phys.* **37**, 1380–1383, (1998).

O'Brien, S., Lang, R., Parker, R., Welch, D.F. and Mehuys, D., *IEEE Photonic Tech. L.* **9**, 440–442, (1997).

Parke, R., Welch, D.F., Hardy, A., Lang, R., Mehuys, D., O'Brian, S., Durko, K. and Scifres, D., *IEEE Photonic Tech. L.* **5**, 297–300, (1993).

Pelton, M., Santori, C., Vuckovic, J., Zhang, B., Solomon, G.S., Plant, J. and Yamamoto, Y., *Phys. Rev. Lett.* **89**, 233602–233605, (2002).

Sagawa, M., Hiramoto, K., Toyonaka, T., Kikiawa, T., Fujisan, S. and Uomi, K., *Electron. Lett.* **32**, 2277–2279, (1996).

Saitoh, T. and Mukai, T., *Coherence, Amplification and Quantum Effect in Semiconductor Lasers*, chapter Traveling-wave semiconductor laser amplifiers. Wiley, New York, (1991).

Sato, S., Osawa, Y. and Saitoh, T., *Jap. J. Appl. Phys.* **36**, 2671–2675, (1997).

Thompson, G.H.B., *Physics of Semiconductor Laser Devices*. Wiley, New York, (1980).

Waks, E., Inoue, K., Santori, C., Fattal, D., Vuckovic, J., Solomon, G.S., Plant, J. and Yamamoto, Y., *Nature* **420**, 762, (2002).

Walpole, J.N., *Opt. Quantum Elect.* **28**, 623–645, (1996).

Willatzen, M., Tanaka, T., Arakawa, Y. and Singh, J., *IEEE J. Quantum Elect.* **30**, 640–653, (1994).

Yariv, A., *Quantum Electronics* (3rd, ed.), J. Wiley, New York, (1989).

Chapter 2

Theory of Quantum Dot
Light–Matter Dynamics

When light 'meets' matter in a semiconductor medium the usual consequence is a 'colourful' interplay of dynamic carrier transport and light propagation involving a broad band of optical frequencies. These fundamental physical processes intimately link dynamics in space and time and are important for the performance of a particular device. To be able to correctly grasp this complexity in the form of appropriate theoretical models and to perform realistic (extensive) numerical simulations then allows for both a fundamental analysis and a prediction of device behaviour. This may then contribute to the development of quantum dot material with improved properties and to the realisation of novel nanostructures and devices with a controlled performance.

In this chapter we introduce a hierarchy of theoretical models that form the basis of the 'numerical experiments' discussed in the next chapters. We here refer to the joint modelling of the dynamics of light and matter. Thereby, the stationary physical materials properties of the constituent materials are considered via effective parameters.

The materials properties as defined by the specific epitaxial structure of an active quantum dot material can be obtained and discussed on the basis of e.g. an 8-band $\mathbf{k} \cdot \mathbf{p}$ (kp) envelope function theory (Cornet $et~al.$, 2006; Tomic $et~al.$, 2006). This theory allows the calculation of the eigenenergies of the carrier system as well as the matrix elements of the respective optical transitions. Moreover, the $\mathbf{k} \cdot \mathbf{p}$ theory allows one to take into account, in particular, the finite potential of the quantum dots. It can be used to calculate dipole matrix elements and energy levels of a given structure. Here, we use the results of these theories for a definition of a parameter set for the simulations. For a more thorough investigation on the level of materials modelling (e.g. on the basis of $\mathbf{k} \cdot \mathbf{p}$ theory or density functional theory or multiband envelope function approach) we refer the reader to the literature (Bester *et al.*, 2006; Boxberg and Tulkki, 2007; Cornet *et al.*, 2006; Grosse *et al.*, 1997; Grundmann *et al.*, 1995; Tomic *et al.*, 2006).

The physics of active photonic nanomaterials generally can be described by many different theories with varying complexity. The theory and computational modelling of the photonics of quantum dot materials and devices ranges from phenomenological modelling on the basis of rate equations to complex quantum-kinetic approaches. In Fig. 2.1 the arrow from bottom to top indicates an increase in

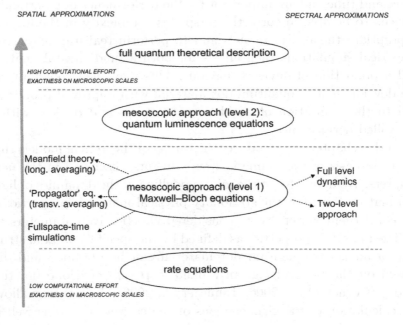

Fig. 2.1.   Hierarchy of model descriptions.

exactness but at the same time an increase in computational effort. Here we will concentrate, in particular, on the dynamic and, indeed, spatio-temporal characteristics of quantum dot nanomaterials and devices. In doing so we can distinguish four levels of description (see Fig. 2.1) with growing levels of complexity:

1. Phenomenological rate equations for the (spatially homogeneous) carrier and photon dynamics.
2. Semiclassical laser theory on the basis of Maxwell–Bloch equations for the spatio-temporal dynamics of the optical fields and the electron-hole plasma (in the figure: level 1 of mesoscopic approaches).
3. Quantum luminescence equations including higher order correlations between light and matter (in the figure: level 2 of mesoscopic approaches).
4. Quantum theoretical description on the basis of equations for the operator dynamics for an analysis of the full light and matter dynamics with consideration of quantum fluctuations.

The first and most simple level of description is based on rate equation models. These models use spatially and temporally averaged quantities for light and matter represented by photon number (or light fields) and carrier densities. They allow the simulation of long-time emission characteristics, analysis of carrier capture, thermionic emission, gain dynamics and output power. Furthermore, the influence of delayed optical feedback can be investigated with suitable extensions to the rate equation approach. This approach is sufficient for the simulation of the behaviour of 'simple' semiconductor lasers, i.e. single-stripe bulk lasers with small (i.e. a few μm) width of the active area that are driven in a regime where nonlinear effects are still negligible (i.e. low power, continuous-wave operation, moderate pump level). Furthermore, the nano-stucture of a medium can only considered via effective macroscopic parameters such as, for example, dot density, refractive index, confinement factors. Inhomogeneities and nonlinear interactions originating from a spatial structuring thus cannot be described.

In many aspects, the Maxwell–Bloch theory (2.) applied to nano-structures represents a compromise between exactness and computational effort. The advantage of this semiclassical approach is the spatially and temporally resolved description of the coupled light–matter dynamics. This theory includes, in particular, the space-dependent material properties typical for 'real' laser systems and the consideration of all relevant nonlinear effects with their respective characteristic interactions with length and time scales ranging from femtoseconds up to nanoseconds. This semiclassical laser theory consequently is most appropriate for a realistic description of physical properties and dynamics of spatially extended semiconductor lasers. In particular, the QD Maxwell–Bloch equations that self-consistently include quantum dot material properties within the framework of a Maxwell–Bloch approach allow a realistic theoretical treatment of nonlinear gain and induced refractive index of quantum dot devices which is of high relevance for many applications. They allow the microscopic calculation of homogeneous and inhomogeneous contributions to the gain and induced index dispersion. They permit a detailed interpretation of relevant physical effects such as saturation, gain build-up and gain recovery or chirp.

A next step in the extensions towards a higher degree in exactness is represented by the quantum luminescence equations. Being based on spatially and temporally resolved equations for light–light and light–matter correlations and using a parameter set for the inclusion of quantum dot material properties these equations still belong to the mesoscopic theories. The inclusion of higher order correlations leads to a higher computational effort but allow for a simulation of quantum coherence properties relevant for the transfer of storage of quantum information.

Finally, the last level of description, the full quantum theoretical approach, considers the dynamics of all operators relevant in the interplay of light and matter.

Generally, all theories have to include a description of the active semiconductor medium and the light field and a description of the macroscopic external constraints imposed by a specific geometry or configuration. This allows the self-consistent inclusion of the complex interplay of microscopic material with macroscopic waveguide and device properties.

Following a brief overview of rate equations we will focus, in particular, on the mesoscopic level of approach and discuss the application of the mesoscopic models (Maxwell–Bloch and quantum luminescence equations) to explore the properties of quantum dot nanomaterials and related devices. This will be the theoretical basis for the work presented in the following chapters.

2.1 Rate Equations

For applications such as frequency conversion and modulated laser systems it is, in particular, the long-time emission behaviour such as relaxation oscillations, dynamic instabilities and modulation response that is of interest. In these cases, it is very common to use phenomenological rate equation models for the simulation of the dynamics of light and matter. Rate equation models (Deppe *et al.*, 1999; Grosse *et al.*, 1997; Huang and Deppe, 2001; Huyet *et al.*, 2004; Jiang and Singh, 1999; Sugawara *et al.*, 2000) describe the time-dependent averaged carrier distributions. They can give first insight into gain saturation and noise properties of QDSOAs (Bilenca and Eisenstein, 2004; Hofmann and Hess, 1999; Qasaimeh, 2004) and allow a quick simulation of the amplification of ultrashort pulse-trains in QDSOAs (Sugawara *et al.*, 2002; Uskov *et al.*, 2004). Furthermore, the influence of delayed optical feedback can be conveniently investigated with simple extensions to the rate equations.

The description of light may refer to power/power fluxes, photon densities or light fields. The simulation of material properties on the other hand may be based on averaged equations for carrier densities or use separate equations for the electron and hole plasma densities.

The rate equation model developed by Sugawara (Sugawara *et al.*, 1997; 2000) which we will describe in the following models the carrier dynamics in the three different layers of the active region, the carrier density in the separate confinement heterostructure (SCH) (N_s), the carrier density in the wetting layer (N_q) and total carrier density in the QD region (N_n). S_m is the photon number of the m-th mode, where $m = 0, 1, \ldots, 2M$ and m corresponds to the mode at E_{cv}. The dot ensemble is divided into $2M + 1$ groups, depending on their resonant energy for the interband transition. The energy width of each group is taken as ΔE. The energy of the n-th group is $E_n = E_{cv} - (M - n)\Delta E$, where $n = 0, 1, 2, \ldots, 2M$ and the dot density is $N_d G_n = N_D G(E_n - E_{cv})\Delta E$, where G represents the energy fluctuation of the dots (typically approximated with a Gaussian distribution function). Longitudinal cavity photon modes with mode separation $\Delta E = ch/(2n_r L_{ca})$ can be included (c is the speed of light in a vacuum, h is Planck's constant, and L_{ca} is the cavity length). They allow the inclusion of the interaction between the dots with different resonant energies. The carrier number in the nth-group QDs is N_n. Then $N = \sum_n N_n$ and the occupation probability in the QD ground state is given as $P_n = N_n/(2N_d V_a G_n)$, where V_A is the active region volume. These rate equations according to the Sugawara model can then be written as

$$\frac{dN_s}{dt} = \frac{I}{e} - \frac{N_s}{\tau_s} - \frac{N_s}{\tau_{sr}} + \frac{N_q}{\tau_{qe}},$$

$$\frac{dN_q}{dt} = \frac{N_s}{\tau_s} + \sum_n \frac{N_n}{\tau_e} - \frac{N_q}{\tau_{qr}} - \frac{N_q}{\tau_{qe}} - \frac{N_q}{\bar{\tau}_d},$$

$$\frac{dN_n}{dt} = \frac{N_q G_n}{\tau_{dn}} - \frac{N_n}{\tau_r} - \frac{N_n}{\tau_e} - \frac{c\Gamma}{n_r} \sum_m g_{mn}^{(1)} S_m,$$

$$\frac{dS_m}{dt} = \frac{\beta N_m}{\tau_r} + \frac{c\Gamma}{n_r} \sum_n g_{mn}^{(1)} S_m - \frac{S}{\tau_p}. \tag{2.1}$$

The Sugawara model includes the relaxation of carriers into the QD ground state. Only a single discrete electron and hole ground state is formed inside the quantum dot and a common time constant is used for various processes like diffusion, relaxation, and recombination. Furthermore, the model assumes charge neutrality for each dot. The model considers the diffusion of the injected carriers through the separate confinement heterostructure (SCH) layer, relaxation into the wetting layer, and relaxation into the dots. Thereby, both radiative and non-radiative recombination processes outside and inside the dots are considered. Some carriers in the ground state lead to stimulated emission contributing to the lasing mode. The associated time constants are diffusion in the SCH region (τ_s), carrier recombination in the SCH region (τ_{sr}), carrier re-excitation from the quantum well to the SCH region (τ_{qe}), carrier re-excitation from the quantum dot to the wetting layer (τ_e), carrier recombination in the wetting layer (τ_e), carrier relaxation into quantum dot (τ_d), and recombination in the quantum dot (τ_r).

β is the spontaneous-emission coupling efficiency to the lasing mode, I is the injected current, $g_{mn}^{(1)}$ is the linear optical gain coupling the mth group dots to the mth mode of the photons (P_{cv} is the transition matrix element):

$$g_{mn}^{(1)} = \frac{2\pi e^2 \hbar N_D}{c n_r \epsilon_0 m_0^2} \frac{|P_{cv}|^2}{E_{cv}} (2P_n - 1) G_n B_{cv}(E_m - E_n). \qquad (2.2)$$

Thereby, the nonlinear optical gain is neglected. Γ is the optical confinement factor and the average carrier relaxation lifetime of τ_d is given by

$$\tau_d^{-1} = \sum_n \tau_{dn}^{-1} G_n = \sum_n \tau_{d0}^{-1}(1 - P_n)G_n \qquad (2.3)$$

where τ_{d0}^{-1} is the relaxation rate for $P_n = 0$, i.e. unoccupied ground state. The photon lifetime in the cavity is given by

$$\tau_p^{-1} = \frac{c}{n_r}[\alpha_i + \ln(1/(R_1 R_2)/(2L_{ca})], \qquad (2.4)$$

where R_1 and R_2 are the cavity mirror reflectivities, The power of the mth mode emitted at one cavity mirror (R_i) is $I_m = \hbar\omega_m c S_m \ln(1/R_i)/(2Ln_r)$.

Similar rate equation models use dynamic equations for the electric field instead of the photon number. This allows the self-consistent inclusion of e.g. phase-sensitive optical feedback (Huyet et al., 2004):

$$\frac{\partial E}{\partial t} = -\frac{E}{2\tau_s} + \frac{\Gamma g_0 N_d}{d}(2\rho - 1)E + i\frac{\delta\omega}{2}E,$$

$$\frac{\partial \rho}{\partial t} = -\frac{\partial \rho}{\partial \tau_d} - g_0(2\rho - 1)|E|^2 + 2R_{cap}(1 - \rho) + R_{esc}\rho,$$

$$\frac{\partial N}{\partial t} = J - \frac{N}{\tau_n} - 2N_d R_{cap}(1 - \rho) - R_{esc}\rho. \tag{2.5}$$

N is the carrier density in the well and ρ is the occupation probability in a dot. τ_s is the photon lifetime, τ_n and τ_d are the carrier lifetime in the well and the dot, respectively. N_d is the two-dimensional density of dots and J denotes the pump term. The factor g_0 is given by the product of cross section of the interaction of photons with the dot carriers and group velocity, $g_0 = \sigma_0 v_{gr}$. Γ is the confinement factor and d is the thickness of the dot layer. The exchange of carriers between the embedding quantum well material and the dots are included via the last two terms in the last line of Eq. (4.2): the carrier capture rate is included via R_{cap} while the carrier escape is described by R_{esc}. The variation of the laser frequency with the carrier densities in the well and in the dots is included via $\delta\omega = \beta_N + \beta_2\rho$, where the coefficient β_1 describes the plasma effect from the carriers in the well, while β_2 describes the variations due to the population in the dots.

Rate equation models for averaged densities are suitable for a description of many long-time effects such as relaxation oscillations or feedback. However, they fail to describe complex carrier effects such as intra-dot scattering between various dot levels via carrier–phonon scattering leading to a complex multi-level dynamics affecting the short time dynamics of quantum dot systems.

As an alternative approach theories on the basis of master equations (Grundmann and Bimberg, 1997a; 1997b; Grundmann *et al.*, 1997) were developed in which the QD ensemble is constituted of microstates, i.e. of subsystems with identical electron-hole densities. This description does not rely on the averaging of the carrier distributions. Instead, the master equation model uses separate equations for electrons and holes and allows for a more detailed insight into dynamic carrier capture effects. Radiative transitions occur between electrons and holes with the same level numbers resulting in photons of energy E_n. Thereby a radiative lifetime τ_r is assumed for all radiative transitions. The filling of the wetting layer and barrier according to external excitation is taken into account via a generation rate (G). From this reservoir carriers are then captured into the dots. Additionally, carriers in the reservoir can recombine with a radiative lifetime τ_b. Thereby, only capture of eh-pairs is considered (i.e. the dots are assumed to be neutral). However, extension of the model to include the separate capture of electrons and holes with identical capture times is also possible.

N_R is the number of eh-pairs in the reservoir and τ_c is the time for a transition of an eh-pair from the reservoir into an empty dot (to any of its M levels). The model assumes a linear decrease of capture time with dot filling, $(1 - n/M)^{-1}\tau_c^0/N_R$ and an infinitely fast inter-level energy relaxation. However, an extension of the model to finite relaxation times is straight forward. The reemission of carriers by the dots back into the reservoir is neglected (low temperature limit). Typical realistic time constants used in the master equation model are $\tau_r \approx \tau_b \approx 1$ ns and carrier capture time τ_c of approximately 10 ps. The dot ensemble is described with so-called microstates: N_n^M denotes the number of QDs filled with n eh-pairs. The master equation for N_n^M assuming random populations then read for $0 < n < M$ (Grundmann and Bimberg, 1997a):

$$\frac{dN_n^M}{dt} = \frac{(n+1)N_{n+1}^M}{\tau_r} - \frac{nN_n^M}{\tau_r}$$

$$+ \frac{N_R N_{n-1}^M}{\tau_c^0}\left(1 - \frac{n-1}{M}\right) - \frac{N_R N_n^M}{\tau_c^0}\left(1 - \frac{n}{M}\right) = 0. \quad (2.6)$$

For $n = 0$ and $n = M$ the corresponding equations read:

$$\frac{dN_0^M}{dt} = \frac{N_1^M}{\tau_r} - \frac{N_R N_0^M}{\tau_c^0} = 0,$$

$$\frac{dN_M^M}{dt} = -\frac{M N_M^M}{\tau_r} + \frac{N_R N_{M-1}^M}{\tau_c^0}\left(1 - \frac{M-1}{M}\right). \quad (2.7)$$

The density in the carrier reservoir is described by

$$\frac{dN_R}{dt} = G - \frac{N_R}{\tau_b} - \frac{N_R}{\tau_c^0}\sum_{n=0}^{M} N_n^M\left(1 - \frac{n}{M}\right) = 0. \quad (2.8)$$

The rate equation and master equation approach represent a very appropriate means for a quick simulation of long-time emission effects and calculation of carrier capture effects. However, they refer to spatially averaged quantities and thus cannot be used for an analysis of the complex space-time dynamics of active nano-structures. They thus cannot describe, for example, beam quality or emission spectra which both depend on the microscopic dynamics in the multi-level charge carrier plasma as well as on the dynamic spatially varying coupling of light and matter.

The focus of the simulation results shown in this book lies in the analysis of space-time effects and interactions. We will thus base the results presented in the following chapters on the mesoscopic approach.

2.2 Maxwell–Bloch Equations

Generally, the Maxwell–Bloch equations (8) consider the dynamics of light fields (via Maxwell's wave equation), the dynamics of the charge carrier plasma (via Bloch equations) and the dynamic interaction between light and matter. This approach thus is based on the semiclassical laser theory, i.e. light is described by classical light fields whereas the dynamics in the active semiconductor is described on a quantum-mechanical level. Relevant dynamic effects such as diffraction, self-focusing, dynamic local carrier generation, carrier recombination by stimulated emission and scattering that determine on a fundamental level the materials properties and device performance are self-consistently included.

For many applications such as the exploration of long-time emission characteristics, the simulation of modulated high-speed devices or the study of beam quality it is sufficient to reduce the active semiconductor medium to a two-level system (two-level approach). The integration of the resulting equations in space and time then allows for a rather efficient simulation of even longer time scales within a reasonable computational effort without too strong losses in the exactness. More detailed models include the full dynamics within the multi-level dots allowing for a very exact description of the dynamic population and depletion of the individual dot levels relevant for e.g. the generation and amplification of ultrashort pulses. Depending on the geometrical constraints given by the width and length of the active nano system the dynamics of the light fields and carriers and be averaged in longitudinal direction (i.e. in direction of the laser cavity, 'Meanfield theory'), averaged in transverse direction (i.e. over the width of the active area, 'Propagator equations') or considered in their full longitudinal and transverse dependence (full space-time simulation).

2.2.1 *Mesoscopic two-level approach*

In the two-level approach the Maxwell–Bloch equations use a two-level Bloch description for the active semiconductor medium. The Bloch equations for a two level homogeneously broadened medium model the complex material polarisation and carrier density interacting with the propagating optical fields. On this level of approach the dynamics of the light fields may either be described by a single-mode wave equation or by so-called multi-mode wave equations including the coexistence and interplay of various longitudinal modes. The dynamic spatio-temporal interplay of longitudinal modes is particular important for the long-time emission dynamics (relaxation oscillations, dynamic instabilities and self-pulsations). In the following, we will apply the multi-mode approach (which generally includes the single-mode operation as a limiting case). These equations represent a very practicable means for the modelling of output power and spectra on long-time scales (i.e. milliseconds) with realistic computational effort. They allow the simulation of the spatio-temporally varying mode competition that is of high importance for the complex interplay of ultrafast light field and carrier dynamics characterising the high-speed dynamics of active semiconductor media (Chow *et al.*, 1994). The simulation results can then be used for a comparison of devices with different properties of the active medium and for a systematic parameter variation in e.g. modulated laser systems.

In order to theoretically describe and simulate these dynamic propagation effects and spatio-temporally varying mode competition we use a multi-mode ansatz for the light fields. The multi-mode Maxwell–Bloch equations consist of multi-mode wave equations and a two-level Bloch description for the active semiconductor medium. The equations for the light fields include the dynamics and coexistence of many longitudinal modes via a mode expansion of the light fields. The transverse dynamics around each longitudinal mode is automatically included within the frame of the wave equations. The resulting multimode Maxwell–Bloch equations consist of multimode wave equations and two-level Bloch equations for the dynamics of carriers as well as the dipole dynamics within the active nano-media.

The spatially dependent multi-mode wave equations for the dynamics of the light fields propagating in forward ('+') and backward ('−') directions read

$$\frac{\partial}{\partial t}E^{\pm} + \frac{\partial}{\partial z}E^{\pm} = iD_p\frac{\partial^2}{\partial x^2}E^{\pm} - i\eta E^{\pm} + \Gamma P^{\pm}_{(0)}. \tag{2.9}$$

The diffraction coefficient is $D_p = (2n_l k_0)^{-1}$ with the vacuum wavenumber $k_0 = 2\pi/\lambda$. Waveguiding properties are included in the parameter η; Γ is the confinement factor. The polarisation P couples the light field dynamics to the dynamics of the carriers within the active medium. On the basis of an effective two-level description of the material properties of the active dot medium the dynamics of the carrier density and the polarisation can be described by the following Bloch equations

$$\frac{\partial}{\partial t}P^{\pm}_{(0)} = -\gamma_p\left[\left(1+i\frac{\bar{\omega}}{\gamma_p}\right) + (\rho+i\sigma)N\right]P^{\pm}_{(0)}$$
$$+ \beta\left((N_{(0)}+i\alpha)E^{\pm} + N_{(1)}E^{\mp}\right)$$

$$\frac{\partial}{\partial t}P^{\pm}_{(1)} = -\gamma_p\left[\left(1+i\frac{\bar{\omega}}{\gamma_p}\right) + (\rho+i\sigma)N\right]P^{\pm}_{(1)} + \beta N_{(1)}E^{\pm}$$

$$\frac{\partial}{\partial t}N_{(0)} = \Lambda + D_f\nabla^2 N_{(0)} - \gamma_{nr}N_{(0)}$$
$$- 2\left(E^{+}\left(P^{+}_{(0)} - \Lambda_0 E^{+}\right)^{*} + E^{-}\left(P^{-}_{(0)} - \Lambda_0 E^{-}\right)^{*} + c.c.\right)$$

$$\frac{\partial}{\partial t} N_{(1)} = -4 D_f k_z^2 N_{(1)} - \gamma_{nr} N_{(1)}$$

$$- 2 \left(E^+ \left(P_{(0)}^- - \Lambda_0 E^- \right)^* + E^{-*} \left(P_{(0)}^+ - \Lambda_0 E^+ \right) \right.$$

$$\left. + E^{+*} P_{(1)}^+ + E^- P_{(1)}^{-*} \right). \tag{2.10}$$

In Eq. (2.9) and (2.10) the parameters $P_{(0)}^\pm$, $P_{(1)}^\pm$, $N_{(0)}$ and $N_{(1)}$ are the (lowest and first order) coefficients of the mode-expansion. Λ describes the carrier injection into the stripes, D_f is the carrier diffusion constant and k_z denotes the wave number of the propagating light fields. $\bar{\omega}$ is the frequency detuning between the frequency of the electron-hole pair and the light frequency. γ_{nr} describes the rate of nonradiative recombination and γ_p is the dephasing of the dipole. The dimensionless constant β determines the maximum gain. The material parameters ρ and σ consider the increase in the polarisation decay rate and the drift of the gain maximum with increasing carrier density, respectively. Being derived from microscopic calculations (Gehrig and Hess, 2003) they include the individual material properties of the quantum-dot device. The α-factor describes the amplitude phase coupling. The parameter Λ_0 guarantees a vanishing gain at transparency. One should note here that the Bloch equations can generally be applied to arbitrary types of active semiconductor material systems (e.g. bulk media, quantum well etc.) and the specific material properties are then included in the respective parameters and functional dependences. Application to a quantum dot ensemble embedded in a quantum well material requires some modifications and extensions in the general Bloch equation in order to guarantee a realistic inclusion of the characteristic properties of a quantum dot ensemble. These modifications are: firstly, an additional term $\frac{\partial N_0}{\partial t}\big|$ describing carrier escape and carrier capture into the dot from the wetting layer (Gehrig and Hess, 2003) must be included in the equation for the carrier density. This term models the dynamic carrier escape and carrier capture from the wetting layer states. The dynamics of the carriers in the layers surrounding the dots thereby is described by a diffusion equation (Gehrig and Hess, 2003). Secondly, in a given quantum dot ensemble $P_{(0,1)}^{+-}(i,j)$

and $\bar{\omega}$ refer to 'interlevel-dipoles' and frequencies of the respective electron and hole levels (with level index i for electrons and j for holes, respectively). The reduced amplitude-phase coupling that is typical for quantum dot lasers, enters the equations via a value of $1 - 1.5$ for the α factor. We note here that for bulk or quantum well medium this value would change to $2.4 - 3$. We would like to note further that the amplitude phase generally varies in space and time. This is fully taken into account in the mesoscopic Maxwell–Bloch description of multi-level systems (see below).

In order to include the superposition and interaction of longitudinal modes in our theoretical description of a spatially extended laser structure we express the light fields, the polarisation and the carrier density in terms of a multi-mode expansion

$$E(z,t) = e^{ik_0z} \sum_{n=0}^{\infty} E_n^+(z,t)e^{2ink_0z} + e^{-ik_0z} \sum_{n=0}^{\infty} E_n^-(z,t)e^{-2ink_0z}$$

$$(2.11)$$

$$P(z,t) = e^{ik_0z} \sum_{n=0}^{\infty} P_n^+(z,t)e^{2ink_0z} + e^{-ik_0z} \sum_{n=0}^{\infty} P_n^-(z,t)e^{-2ink_0z}$$

$$(2.12)$$

$$N(z,t) = N_{(0)}(z,t) + \sum_{n=0}^{\infty} N_n(z,t)e^{2ink_0z} + \text{c.c.} \qquad (2.13)$$

Insertion of this expansion in the Maxwell–Bloch equations (Mnkel *et al.*, 1996) leads to a hierarchy of equations for the polarisation P and the carrier density N. Carrier diffusion and light diffraction lead to a smoothing of the grating structure induced in and by the counterpropagating light fields. In addition the spatio-temporal light field dynamics couples via the polarisiation in the wave equation to the full (nonlinear) semiconductor properties and thus to the counterpropagating waves at every grid point of the laser. As a consequence, we only have to consider the lowest order in the equations.

This leads to the following set of equations for the dynamics of the light fields propagating in the forward ('+') and backward ('−')

directions in the laser

$$\frac{\partial}{\partial t}E^{\pm} + \frac{\partial}{\partial z}E^{\pm} = \mathrm{i}D_{\mathrm{p}}\frac{\partial^2}{\partial x^2}E^{\pm} - \mathrm{i}\eta E^{\pm} - \mathrm{i}k_0 E^{\mp} + \beta\Gamma P^{\pm}_{(0)}. \quad (2.14)$$

The diffraction coefficient is $D_{\mathrm{p}} = (2n_1 k_0)^{-1}$ with the vacuum wavenumber $k_0 = 2\pi/\lambda$. The wave-guiding properties derived from the effective index approximation are included in the parameter η. $\bar{\omega}$ denotes the frequency detuning between the frequency of the electron-hole pair and the light frequency. Via the polarisation the light fields are locally coupled to carriers in the active medium.

The time-dependent calculation of the light field and carrier dynamics on the basis of the set of Eqs. (2.14) and (2.10) allows for a simulation of the long-time behaviour of spatially extended quantum dot lasers with explicit consideration of the interplay of a number of longitudinal modes. This effective multi-mode theory will in the following chapters be applied in such cases where we aim at the simulation of long-time emission dynamics of semiconductor lasers.

2.2.2 *Mesoscopic Maxwell–Bloch description of multi-level quantum dot systems*

The derivation of Maxwell–Bloch equations for multi-level QD systems allows a fundamental analysis of the complex level dynamics in the charge carrier plasma and of the spatially and temporally varying interplay of light and matter. They can be used to analyse the microscopic carrier dynamics in a quantum dot and to derive relevant time scales for e.g. carrier capture and escape. These QD Maxwell Bloch equations (Gehrig and Hess, 2003) describe the spatio-temporal light field and inter-/intra-level carrier dynamics in each QD of a typical QD-ensemble in quantum dot lasers. The theory includes spontaneous luminescence, counterpropagation of amplified spontaneous emission and induced recombination as well as carrier diffusion in the wetting layer of the quantum dot laser. Intra-dot scattering via emission and absorption of phonons, as well as the scattering with the carriers and phonons of the surrounding wetting layer are dynamically included on a mesoscopic level. Typical spatial fluctuations in size and energy levels of the quantum dots and irregularities in the spatial distribution of the quantum dots in the active layer are simulated via statistical methods.

In the following we give an overview of the microscopic (i.e. frequency resolved) approach on the basis of semiclassical semiconductor laser theory. The theory was originally developed for bulk semiconductor media such as laser arrays, broad-area lasers or VCSELs (Hess, 1993; Grundmann *et al.*, 1995) and has then be extended to describe quantum well (Hamm *et al.*, 2002) and quantum dot (Gehrig and Hess, 1998) media. In contrast to both rate equation models and the two-level Maxwell–Bloch description, the QD Maxwell–Bloch equations for multi-level systems include the full space and momentum dependence of the charge carrier distributions and the polarisation.

The model will be set up in a very general way and thus principally hold for variable material systems. Without loss of generality, we will in the simulations concentrate on the InGaAs/GaAs system. Macroscopic device properties such as laser geometry, electronic contacts, epitaxial structure, wave-guiding properties, mirror design, resonator shape or operation conditions are fully taken into account. The model generally allows an investigation of a multitude of quantum dot systems such as large-area lasers and amplifiers, laser arrays and lasers with delayed optical feedback (i.e. embedded in an external resonator). In the multi-level semiclassical approach the optical fields and the polarisation are described in a classic manner whereas the dynamics of the carrier distribution and the inter-level polarisation are considered quantum-mechanically. We will derive a set of equations consisting of wave equations for the propagation of light fields in the spatially extended resonator of the semiconductor laser and the semiconductor Bloch equations for the microscopic dynamics of the charge carrier plasma in the energy levels of the active quantum dot ensemble.

2.2.2.1 *Optical field dynamics*

Starting with Maxwell's equations one can derive the following wave equation for the optical fields $\vec{E}$ in a dielectric medium (Gehrig and

Hess, 1998)

$$\frac{1}{\varepsilon_0}\nabla\nabla\cdot\vec{P} + \nabla^2\vec{E} - \frac{1}{c^2}\frac{\partial^2}{\partial t^2}\vec{E} = \mu_0\frac{\partial^2}{\partial t^2}\vec{P}. \qquad (2.15)$$

Since typical geometries of the active area favour the propagation of light fields perpendicular to the facets (in the following the z-coordinate) it is a common method to devide the fields and operators into longitudinal and transverse parts. Furthermore, we separate the nonlinear part of the polarisation ($\vec{P}_{\mathrm{nl}}$) according to

$$\vec{P} = \vec{P}_1 + \vec{P}_{\mathrm{nl}} = \varepsilon_0\chi_1\vec{E} + \vec{P}_{\mathrm{nl}}. \qquad (2.16)$$

Expanding the wave equations in powers of the dimensionless number $f = \frac{w}{l} = \frac{1}{k_z w} \ll 1$ and keeping terms and equations up to first order finally leads to

$$\pm\frac{\partial}{\partial z}\vec{E}(\vec{r},t) + \frac{n_1}{c}\frac{\partial}{\partial t}\vec{E}(\vec{r},t)$$
$$= \frac{i}{2k_z}\frac{\partial^2}{\partial x^2}\vec{E}(\vec{r},t) - \left(\frac{\alpha}{2} + i\eta\right)\vec{E}(\vec{r},t) + \frac{i}{\varepsilon_0}\frac{k_z}{2n_1^2}\vec{P}_{\mathrm{nl}}(\vec{r},t). \qquad (2.17)$$

In first order Maxwell's wave equation for the optical fields $\vec{E}(\vec{r},t)$ and the polarisation $\vec{P}_{\mathrm{nl}}(\vec{r},t)$ is purely transverse, but nevertheless includes the full transverse and longitudinal dependence of all quantities.

α is the linear absorption and η includes static and dynamic changes in the permittivity affecting the refractive index and the propagation wavenumber. The spatial dependence of the static waveguiding structure resulting from the lateral and vertical confinement of the active area can be calculated on the basis static perturbation leading to effective parameters such as the background refractive index n_1 of (2.17). The position vector $\vec{r} = (x,z)$ denotes the lateral (x) and longitudinal (z) direction, k_z denotes the (unperturbed) wavenumber of the propagating fields. V is the normalisation volume of the crystal; n_1 is the background refractive index of the active layer.

Depending on the epitaxial growth process the laser may consist of several layers defining vertical 'QD stacks' (columns). The layered structuring in vertical direction (y) is considered via effective material and device parameters. In particular, these are the effective refractive index and the guiding properties of the layer as well as the physical properties of the QD stack (vertically averaged energy levels, damping rates or QD size). The vertically averaged physical properties characterise an 'effective' QD. The properties that enter the equations in a self-consistent way are the energy levels, the initial occupation of the levels (established e.g. via optical or electrical pumping) and the size of the QDs.

In many cases the quantum dot ensemble of a device is arranged in an in-plane configuration. Figure 2.2 schematically shows a typical geometry: the active layer contains an ensemble of spatially distributed QDs that are embedded in the quantum well wetting layer (WL). Light propagates within the active layer in the resonator predominately along the longitudinal (z) direction. The active area then acts as a Fabry–Perot resonator for the counterpropagating light fields. This requires the explicit consideration of light propagating in forward ('+') and backward ('−') directions along the axis

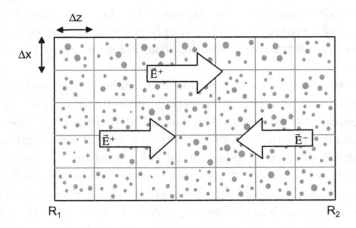

Fig. 2.2.   Representation of the quantum dot laser model geometry. The counter-propagating light fields ($E^{\pm}$) spatio-temporally couple with carriers in the ensemble of quantum dots. Characteristic fluctuations in size and location of the quantum dots are effectively represented on a numerical grid with equally spaced grid points in the lateral (x) and propagation (z) direction.

z-direction) of the resonator leading to the following wave equation:

$$\pm \frac{\partial}{\partial z}\vec{E}^{\pm}(\vec{r},t) + \frac{n_{\mathrm{l}}}{c}\frac{\partial}{\partial t}\vec{E}^{\pm}(\vec{r},t)$$

$$= \frac{\mathrm{i}}{2k_z}\frac{\partial^2}{\partial x^2}\vec{E}^{\pm}(\vec{r},t) - \left(\frac{\alpha}{2} + \mathrm{i}\eta\right)\vec{E}^{\pm}(\vec{r},t) + \frac{\mathrm{i}}{\epsilon_0}\frac{k_z}{2n_{\mathrm{l}}^2}\vec{P}_{\mathrm{nl}}^{\pm}(\vec{r},t), \quad (2.18)$$

where

$$\vec{P}_{\mathrm{nl}}^{\pm}(\vec{r},t) = 2V^{-1}\sum_{i,j}d(j,i)p_{\mathrm{nl}}^{\pm}(i,j,\vec{r},t) \qquad (2.19)$$

is the polarisation which is composed of microscopic inter-level dipoles. Those dipoles are the result of the interaction of the electron and hole levels of the dot ensemble. i and j denote the index for the electron and hole levels, respectively, $d(j,i)$ is the dipole matrix element for a transition between level i and j. Thereby, every combination of electron and hole states is considered in the multi-level dot system. Via the polarisation of the active QD medium, the light fields are locally coupled to the dynamics of the carriers and to the inter-level dipole dynamics.

The light propagating in the longitudinal direction is partially reflected and partially transmitted, depending on the device reflectivities, R_1 and R_2, leading to the following boundary conditions

$$\vec{E}^{+}(x, z = 0, t) = -\sqrt{R_1}\vec{E}^{-}(x, z = 0, t),$$

$$\vec{E}^{-}(x, z = L, t) = -\sqrt{R_2}\vec{E}^{+}(x, z = L, t), \qquad (2.20)$$

where L denotes the length of the device. R_1 and R_2 may be the natural reflectivities given by the material-air boundaries or represent effective facet reflectivities defined by e.g. a Bragg mirror (as used in VCSELs) or antireflection coating (as required in laser amplifiers). At the lateral edges, the active area is surrounded by (un-pumped) layers of semiconductor media which strongly absorb the fields (absorption coefficient α_{w}). This is represented by the lateral boundary conditions

$$\frac{\partial \vec{E}^{\pm}}{\partial x} = -\alpha_{\mathrm{w}}\vec{E}^{\pm} \qquad (2.21)$$

at $x = w/2$ (w is the full lateral width of the laser structure) as well as

$$\frac{\partial \vec{E}^{\pm}}{\partial x} = +\alpha_{\mathrm{w}} \vec{E}^{\pm} \tag{2.22}$$

at $x = -w/2$. Via the polarisation of the active QD medium, the light fields are locally coupled to the dynamics of the carriers and to the inter-level dipole dynamics which will be discussed next.

2.2.2.2 Carrier dynamics within a quantum dot

Starting from the single particle density matrices for the electrons, $n^{\mathrm{e}} = \langle c^{\dagger}c \rangle$, and holes, $n^{\mathrm{h}} = \langle d^{\dagger}d \rangle$, and for the inter-level polarisation, $p = \langle d^{\dagger}c \rangle$, where c and d are the local annihilation operators for electrons and holes respectively, one can derive semiconductor Bloch equations specifically for quantum dots. The resulting quantum dot multi-level Bloch equations mesoscopically describe the dynamic changes of the electron and hole distributions inside the dot (for each energy level) and the dynamics of the (inter-level) dipoles. If one considers an ensemble of quantum dots as the active medium in a quantum dot laser additional terms and effects are of relevance. These are contributions describing the electrical injection of carriers (pumping) Λ^{e} (including Pauli-blocking), induced recombination (with generation rate $g^{\mathrm{e,h}}$), spontaneous recombination of the carriers (Γ_{sp}), carrier–carrier and carrier–phonon scattering for the intra-dot relaxation ($\partial n^{\mathrm{e,h}}/\partial t|_{\mathrm{QD}}^{\mathrm{c-ph}}$) and the interaction with the wetting layer ($\partial n^{\mathrm{e,h}}/\partial t|_{\mathrm{QD-WL}}$). The dynamics of the occupation of electrons (e, level index i) and holes (h, level index j), $n^{\mathrm{e,h}}$, and the dynamics of the inter-level polarisations $p^{\pm}$ (coupled to the forward ('+') and backward ('−') propagating optical fields) within a QD are then governed by the equations of motion

$$\frac{\partial n^{\mathrm{e}}(i)}{\partial t} = \Lambda^{\mathrm{e}}(i)\left(D^{\mathrm{e}}(i) - n^{\mathrm{e}}(i)\right) + g^{\mathrm{e}}(i) - \gamma_{\mathrm{nr}}n^{\mathrm{e}}(i)$$

$$- \sum_{j} \Gamma_{\mathrm{sp}}n^{\mathrm{e}}(i) \cdot n^{\mathrm{h}}(j) + \left.\frac{\partial n^{\mathrm{e}}(i)}{\partial t}\right|_{\mathrm{QD}}^{\mathrm{c-ph}} + \left.\frac{\partial n^{\mathrm{e}}(i)}{\partial t}\right|_{\mathrm{QD-WL}}$$

$$\frac{\partial n^{\mathrm{h}}(j)}{\partial t} = \Lambda^{\mathrm{h}}(j)\left(D^{\mathrm{h}}(j) - n^{\mathrm{h}}(j)\right) + g^{\mathrm{h}}(j) - \gamma_{\mathrm{nr}}n^{\mathrm{e}}(i)$$

$$- \sum_i \Gamma_{\mathrm{sp}} n^{\mathrm{h}}(j) \cdot n^{\mathrm{e}}(i) + \left.\frac{\partial n^{\mathrm{e}}(i)}{\partial t}\right|_{\mathrm{QD}}^{\mathrm{c-ph}} + \left.\frac{\partial n^{\mathrm{e}}(i)}{\partial t}\right|_{\mathrm{QD-WL}}$$

$$\frac{\partial p^{\pm}(j,i)}{\partial t} = -(\mathrm{i}\bar{\omega}(j,i) + \gamma_{\mathrm{p}})p^{\pm}(j,i) - \frac{\mathrm{i}}{\hbar}\left[n^{\mathrm{e}}(i) + n^{\mathrm{h}}(j)\right]\mathcal{U}^{\pm}$$

$$- \frac{\mathrm{i}}{\hbar}\delta\mathcal{U}_{\mathrm{nl}}^{\pm} + F_{\mathrm{p}}q^{\mathrm{p}} + \left.\frac{\partial p^{\pm}(j,i)}{\partial t}\right|_{\mathrm{QD}}^{\mathrm{p-ph}} \qquad (2.23)$$

where γ_{nr} represents the rate due to nonradiative recombination and τ_{p} denotes the dephasing time of the inter-level dipole. The pump term

$$\Lambda^{\mathrm{c}}(l) = \Gamma_{\mathrm{QDS}}\frac{I\eta}{eh}\frac{n_{\mathrm{eq}}^{\mathrm{c}}(l)}{\sum_l n_{\mathrm{eq}}^{\mathrm{c}}(l)\left(D^{\mathrm{e,h}}(l) - n^{\mathrm{c}}(l)\right)} \qquad (2.24)$$

mesoscopically represents the carrier injection and includes the pump-blocking effect (c = e, h and $l = i, j$ for electrons and holes respectively). It depends on the absolute injection current, I, pump efficiency η, and the thickness of the active area, h. $D^{\mathrm{c}}(l)$ denotes the degeneracy of an end energy level (i.e. the maximum occupation with carriers). Γ_{QDS} describes the reduction of the pump efficiency resulting from the vertically arranged QDs, i.e. the 'spatial overlap' between carrier injection and a vertical stack of QDs in the medium.

The generation rates given by

$$g^{\mathrm{e}}(i) = \mathrm{Re}\left[\frac{\mathrm{i}}{\hbar}\sum_j [(\mathcal{U}^+ p_{\mathrm{nl}}^{+*}(j,i) + \mathcal{U}^- p_{\mathrm{nl}}^{-*}(j,i))\right.$$

$$\left. - (\mathcal{U}^{+*} p_{\mathrm{nl}}^+(j,i) + \mathcal{U}^{-*} p_{\mathrm{nl}}^-(j,i))]\right]$$

$$g^{\mathrm{h}}(j) = \mathrm{Re}\left[\frac{\mathrm{i}}{\hbar}\sum_i [(\mathcal{U}^+ p_{\mathrm{nl}}^{+*}(j,i) + \mathcal{U}^- p_{\mathrm{nl}}^{-*}(j,i))\right.$$

$$\left. - (\mathcal{U}^{+*} p_{\mathrm{nl}}^+(j,i) + \mathcal{U}^{-*} p_{\mathrm{nl}}^-(j,i))]\right] \qquad (2.25)$$

depend on the inter-level polarisation p and on the optical field contributions of spontaneous and induced emission constituting the local field $\mathcal{U}^\pm$. The Langevin noise term $F_p q^p$ describes dipole fluctuations (Hofmann and Hess, 1999) with amplitude $F_p = \frac{\Gamma\sqrt{2\hbar\epsilon_r}}{n_1^2 L\sqrt{\epsilon_0\omega_0}}$. The local fields $\mathcal{U}^\pm = \vec{d}(j,i)\vec{E}^\pm + \delta\mathcal{U}^\pm$ are composed of the optical light field contributions $E^\pm$ as well as those induced by Coulomb screening in each quantum dot and by the Coulomb interactions between the carriers in the QD and the carriers in the wetting layer, $\delta\mathcal{U}$. $\vec{d}(j,i)$ is the inter-level dipole matrix element. The inter-level polarisation depends via $\bar{\omega}(j,i) = \hbar^{-1}\left(\mathcal{E}^e + \mathcal{E}^h\right) - \omega$ (ω is the frequency of the propagating light fields) on the carrier energies $\mathcal{E}^{e,h}$ that are given by

$$\mathcal{E}^c(l) = \epsilon^c(l) + \delta\mathcal{E}^c(l), \tag{2.26}$$

with the unperturbed level energies $\epsilon^{e,h}$ (i.e. neglecting the carrier dynamics). The characteristic level energies $\epsilon^c(l)$ of the unperturbed QD are taken from microscopic material calculations (Grosse *et al.*, 1997) and are self-consistently included in the theory. The Coulomb-induced screening that leads to a renormalisation of these energy levels and also results in additional local field contributions strongly depends on the specific QD design (size, shape). These respective corrections have been determined in detailed calculations (e.g. Bolcatto and Proetto, 1999; Braskén *et al.*, 2000; Oshiro *et al.*, 1999) and are represented in the quantum dot Maxwell–Bloch equations (QDSBEs) (2.23) in the form of spatially dependent energies ($\hbar\bar{\omega}$ and local field contributions $\delta\mathcal{U}^\pm$).

The QD multi-level Bloch equations allow the fundamental analysis of spatio-spectral hole burning and saturation as well as the influence of spatial effects arising from a spatially varying size and density of the dots. They consider the space-dependent interaction of the carriers in a QD with the optical fields (associated with spontaneous and induced emission processes). In particular, the spatial and spectral characteristics of the medium are fully taken into account in the mesoscopic approach. This includes the localisation of the dots in the medium, fluctuations in size and shape of the

QDs, the spatially dependent light field propagation and diffraction as well as spatially dependent scattering processes and carrier transport.

The time-dependent calculation of the carrier distributions and the light field dynamics allow for explicit consideration of the individual time scales of the various interaction processes. The relevant time scales range from the femtosecond regime (for the fast carrier scattering processes) up to the picosecond and nanosecond regime (for the dynamics of the propagating light fields and of the spatial carrier density). The approach taken here incorporates the full microscopic spatio-temporal dynamics together with the relevant macroscopic properties of typical semiconductor lasers. It generally applies to semiconductor laser systems with a large variety of geometries and active material systems. To be specific, in this book we will focus on InGaAs/GaAs based systems (Adachi, 1985). The relevant material and structural parameters used in the simulations are given in Table 2.1.

Table 2.1 Fundamental material and device parameters of GaAs-based semiconductor lasers.

Parameter	Physical quantity	Value
n_c	Refractive index of the cladding layers (GaAlAs)	3.35
n_l	Refractive index of active layer (GaAs)	3.59
λ	Laser wavelength	815 nm–930 nm
R_1	Front facet mirror reflectivity	$10^{-4} \dots 1$
R_2	Back facet mirror reflectivity	$10^{-4} \dots 1$
τ_{nr}	Nonradiative recombination time	5 ns
a_0	Exciton Bohr radius	1.243×10^{-6} cm
m_0	Mass of the electron	$9.1093879 \times 10^{-31}$ kg
m_e	Effective electron mass	$0.067\, m_0$
m_h	Effective hole mass	$0.246\, m_0$
$\mathcal{E}_g(0)$	Semiconductor energy gap at $T = 0$ K	1.519 eV
η_i	Injection efficiency	0.5
Γ	Confinement factor	0.55/0.54
α_w	Absorption	$30\,\mathrm{cm}^{-1}$
v_{sr}	Surface recombination velocity	10^6 m/s

2.3   Quantum Luminescence Equations

In recent years increasing miniaturisation and high-speed application has led to a growing influence of quantum properties of the light fields on the spatio-temporal dynamics and emission properties of modern nano-structures. A theoretical analysis of quantum effects thus is of high importance for a profound understanding of the physical properties in advanced nano systems. The underlying physical mechanisms that are responsible for the generation of light affect the interplay between spontaneous and induced radiation and consequently the spatial and spectral emission as well as noise properties and loss mechanisms.

The quantum theory used in this book builds on the concepts of Maxwell–Bloch approaches for light–matter interaction processes leading to quantum Maxwell–Bloch equations for nano-structures. These equations are derived from fully quantum mechanical operator dynamics describing the interaction of the light field with the quantum states of the electrons and the holes in the dot. Our approach (Gehrig and Hess, 2008) includes on a microscopic basis the interactions and correlations of the amplitudes of the light fields and the dipoles excited within the medium that cause and mediate the (spatial and temporal) coherence of spontaneous and stimulated emission. Field–field correlations and field–dipole correlations lead to quantum noise effects that cause spontaneous emission and amplified spontaneous emission. These processes are self-consistently taken into account.

The coherence properties of the emitted radiation are directly reflected in the dynamics of the expectation value of the field–field correlation operator $\langle \hat{b}_{\mathbf{R}}^{\dagger} \hat{b}_{\mathbf{R}'} \rangle$ and the expectation value of the field–dipole correlation operator $\langle \hat{b}_{\mathbf{R}}^{\dagger} \hat{c}_{\mathbf{R}'} \hat{d}_{\mathbf{R}''} \rangle$ where $\hat{b}_{\mathbf{R}}$ ($\hat{b}_{\mathbf{R}}^{\dagger}$) is the (spatially dependent) annihilation (creation) operator for photons, $\hat{c}_{\mathbf{R}}$ ($\hat{c}_{\mathbf{R}}^{\dagger}$) and $\hat{d}_{\mathbf{R}}$ ($\hat{d}_{\mathbf{R}}^{\dagger}$) are the annihilation (creation) operators for electrons and holes, respectively. The discrete positions $\mathbf{R}$ correspond to the lattice sites of the Bravais lattice describing the semiconductor crystal. Each lattice site actually represents the spatial volume ν_0 of the Wigner–Seitz cell of the lattice. On the

basis of these operators one can derive equations of motion for the
expectation values of the carriers ($\langle \hat{c}_{\mathbf{R}}^\dagger \hat{c}_{\mathbf{R}'} \rangle$ and $\langle \hat{d}_{\mathbf{R}}^\dagger \hat{d}_{\mathbf{R}'} \rangle$), dipoles
($\langle \hat{c}_{\mathbf{R}} \hat{d}_{\mathbf{R}'} \rangle$), fields ($\langle \hat{b}_{\mathbf{R}} \rangle$) and, in particular, equations for the expectation values of the operators describing field–field ($\langle \hat{b}_{\mathbf{R}}^\dagger \hat{b}_{\mathbf{R}'} \rangle$) and
field–dipole correlations ($\langle \hat{b}_{\mathbf{R}}^\dagger \hat{c}_{\mathbf{R}'} \hat{d}_{\mathbf{R}''} \rangle$) (Deppe *et al.*, 1999). For that
purpose the field–field and field–dipole correlation variables are
defined on a quasi-continuous length scale, $\rho^I(\mathbf{r} = \mathbf{R}; \mathbf{r}' = \mathbf{R}') = \nu_0^{-1} \langle \hat{b}_{\mathbf{R}}^\dagger \hat{b}_{\mathbf{R}'} \rangle$, $\Theta^{corr.}(\mathbf{r} = \mathbf{R}; \mathbf{r}' = \mathbf{R}', \mathbf{r}'' = \mathbf{R}'') = \sqrt{\nu_0^3}^{-1} \langle \hat{b}_{\mathbf{R}}^\dagger \hat{c}_{\mathbf{R}'} \hat{d}_{\mathbf{R}''} \rangle$.
The variables are then transformed into Wigner functions
$I(\mathbf{r}; \mathbf{r}', k_e, k_h) = \int d^3 r' e^{-i k_{\mathbf{eh}} \mathbf{r}'} \rho^I \left(\mathbf{r} - \frac{\mathbf{r}'}{2}, \mathbf{r} + \frac{\mathbf{r}'}{2} \right)$, $C(\mathbf{r}; \mathbf{r}', k_e, k_h) = \int d^3 r'' e^{-i k_{\mathbf{eh}} \mathbf{r}''} \Theta^{corr.} \left(\mathbf{r}; \mathbf{r}' - \frac{\mathbf{r}''}{2}, \mathbf{r}' + \frac{\mathbf{r}''}{2} \right)$ where $k_{\mathbf{eh}}$ is the wave vector
of the electron-hole pair. Using the local approxmation finally leads
to the following equations of motion for the field–field correlation I
and the field–dipole correlation C (Gehrig and Hess, 2008)

$$\frac{\partial}{\partial t} I_{ij}(\mathbf{r}, \mathbf{r}', k_e, k_h) = -(\kappa_i + \kappa_j) I_{ij}(\mathbf{r}, \mathbf{r}', k_e, k_h)$$

$$- i \frac{\omega_0}{2k_0^2 \epsilon_r} \left(\Delta_\mathbf{r} - \Delta_{\mathbf{r}'} \right) I_{ij}(\mathbf{r}, \mathbf{r}', k_e, k_h)$$

$$- i g_0 \frac{\sqrt{\nu_0}}{4\pi^2} \left[C_{ij}(\mathbf{r}, \mathbf{r}', k_e, k_h) - C_{ji}^*(\mathbf{r}', \mathbf{r}, k_e, k_h) \right.$$

$$\left. + \frac{1}{k_0^2 \epsilon_r} \left(\frac{\partial^2}{\partial r_j' r_k'} C_{ik}(\mathbf{r}, \mathbf{r}', k_e, k_h) - \frac{\partial^2}{\partial r_i r_k} C_{jk}^*(\mathbf{r}', \mathbf{r}, k_e, k_h) \right) \right].$$

$$(2.27)$$

$$\frac{\partial}{\partial t} C_{ij}(\mathbf{r}, \mathbf{r}', k_e, k_h) = [-\Gamma_{k_e, k_h} - i\Omega_{k_e, k_h}] C_{ij}(\mathbf{r}, \mathbf{r}', k_e, k_h)$$

$$- (\kappa_i - i\delta\omega_i) C_{ij}(\mathbf{r}, \mathbf{r}', k_e, k_h) - i \frac{\omega_0}{2k_0^2 \epsilon_r} \Delta_\mathbf{r} C_{ij}(\mathbf{r}, \mathbf{r}', k_e, k_h)$$

$$+ i g_0 \sigma \sqrt{\nu_0} \left(f^e(\mathbf{r}', k_e, k_h) \right) + f^h((\mathbf{r}', k_e, k_h) - 1) U_{ij}(\mathbf{r}, \mathbf{r}', k_e, k_h)$$

$$+ i g_0 \sigma \sqrt{\nu_0} \, \delta(\mathbf{r} - \mathbf{r}') \delta_{ij} (f^e(\mathbf{r}, k_e, k_h) \cdot f^h(\mathbf{r}, k_e, k_h)$$

$$+ p(\mathbf{r}, k_e, k_h) \cdot p^*(\mathbf{r}, k_e, k_h)). \qquad (2.28)$$

The subscripts i and j denote the index for the two (x and y)
spatial dimensions, respectively, i.e. $i = 1, \ldots, nx$, $j = 1, \ldots, ny$

where nx and ny are the number of grid points in x and y direction. κ is the cavity loss rate. The birefringence term $\delta\omega$ includes the difference between the bandgap frequency and the longitudinal frequencies of the confined light for the two polarisation directions. $\Delta_{\mathbf{r}} = \sum_k \partial^2/\partial r_k^2$ and the coupling frequency g_0 is the coupling constant characterising the interaction of carriers and photons. σ is the confinement factor in z direction. In Eq. (2.28) the local field $U_{ij}(\mathbf{r}, \mathbf{r}', k_e, k_h)$ consists of the field–field correlations and a term describing the renormalisation due to the Coulomb interactions. It allows the explicit consideration of the modifications induced in the carrier potential by many-body effects. Further details on the general approach can be found in (Deppe *et al.*, 1999). $f^e(\mathbf{r}, k_e, k_h)$, $f^h(\mathbf{r}, k_e, k_h)$ and $p(\mathbf{r}, k_e, k_h)$ are the (dynamically calculated) electron and hole distributions and the inter-level polarisation with k_e and k_h being the wavenumbers of the electron and hole states, respectively. k_0 is vacuum wavevector corresponding to the central frequency ω_0 and ϵ_r is the relative permittivity. $\Gamma_{k_e,k_h} = \left(\Gamma_{k_e,k_h}^{e,out} + \Gamma_{k_e,k_h}^{e,in} + \Gamma_{k_e,k_h}^{h,out} + \Gamma_{k_e,k_h}^{h,in} \right)$ includes in and out scattering processes as discussed in (Deppe *et al.*, 1999). Many particle effects such as band gap renormalisation or the Coulomb enhancement lead to corrections in the carrier energies and carrier potential (Sugawara *et al.*, 2000). The corrections in carrier energies are considered via the momentum dependent frequency Ω_{k_e,k_h}.

The dynamics of the carrier distributions $f^e(\mathbf{r}, k_e, k_h)$, $f^h(\mathbf{r}, k_e, k_h)$ is given by

$$\frac{\partial}{\partial t} f_\sigma^e(\mathbf{r}, k_e, k_h) = \Lambda^e(\mathbf{r}) - \gamma_{nr} f_\sigma^e(\mathbf{r}, k_e, k_h)$$

$$+ i g_0 \frac{\sqrt{\nu_0}}{2\pi} \left(\sum_i (C_{ii}(\mathbf{r}; \mathbf{r}, k_e, k_h) - C_{ii}^*(\mathbf{r}; \mathbf{r}, k_e, k_h)) \right.$$

$$+ \sum_{ij} \frac{1}{k_0^2 \epsilon_r} \int d^2\mathbf{r}' \delta(\mathbf{r} - \mathbf{r}') \left(\frac{\partial^2}{\partial r_i' r_j'} C_{ij}(\mathbf{r}; \mathbf{r}', k_e, k_h) \right.$$

$$\left. \left. - \frac{\partial^2}{\partial r_i r_j} C_{ij}^*(\mathbf{r}'; \mathbf{r}, k_e, k_h) \right) \right). \tag{2.29}$$

$$\frac{\partial}{\partial t} f_\sigma^h(\mathbf{r}, k_e, k_h) = \Lambda^h(\mathbf{r}) - \gamma_{nr} f^h(\mathbf{r}, k_e, k_h)$$

$$+ i g_0 \frac{\sqrt{\nu_0}}{2\pi} \left(\sum_i (C_{ii}(\mathbf{r}; \mathbf{r}, k_e, k_h) - C_{ii}^*(\mathbf{r}; \mathbf{r}, k_e, k_h)) \right.$$

$$+ \sum_{ij} \frac{1}{k_0^2 \epsilon_r} \int d^2 \mathbf{r}' \delta(\mathbf{r} - \mathbf{r}') \left(\frac{\partial^2}{\partial r_i' r_j'} C_{ij}(\mathbf{r}; \mathbf{r}', k_e, k_h) \right.$$

$$\left. \left. - \frac{\partial^2}{\partial r_i r_j} C_{ij}^*(\mathbf{r}'; \mathbf{r}, k_e, k_h) \right) \right). \tag{2.30}$$

The field–dipole correlation obeys

$$\frac{\partial}{\partial t} C_{ij}(\mathbf{r}; \mathbf{r}', k_e, k_h) = \left[- \left(\Gamma_{k_e, k_h}^{e, out} + \Gamma_{k_e, k_h}^{e, in} + \Gamma_{k_e, k_h}^{h, out} + \Gamma_{k_e, k_h}^{h, in} \right) \right.$$

$$\left. - i\Omega(k_e, k_h) \right] \cdot C_{ij}(\mathbf{r}; \mathbf{r}', k_e, k_h)$$

$$- (\kappa_i - i\delta\omega_i) C_{ij}(\mathbf{r}; \mathbf{r}', k_e, k_h) - i\frac{\omega_0}{2k_0^2 \epsilon_r} \Delta_\mathbf{r} C_{ij}(\mathbf{r}; \mathbf{r}', k_e, k_h)$$

$$+ i g_0 \sigma \sqrt{\nu_0} \left(f^e(\mathbf{r}', k_e, k_h) \right) + f^h(\mathbf{r}', k_e, k_h) - 1) U_{ij}(\mathbf{r}; \mathbf{r}')$$

$$+ i g_0 \sigma \sqrt{\nu_0} \, \delta(\mathbf{r} - \mathbf{r}') \delta_{ij} (f^e(\mathbf{r}, k_e, k_h) \cdot f^h(\mathbf{r}, k_e, k_h)$$

$$+ p(\mathbf{r}, k_e, k_h) \cdot p^*(\mathbf{r}, k_e, k_h)), \tag{2.31}$$

with

$$U_{ij}(\mathbf{r}; \mathbf{r}', k_e, k_h) = I_{ij}(\mathbf{r}; \mathbf{r}', k_e, k_h) + \int_{k_e', k_h'} V_{k_e, k_h, k_e', k_h'} I_{ij}(\mathbf{r}; \mathbf{r}', k_e', k_h'), \tag{2.32}$$

and the field–field correlation is governed by

$$\frac{\partial}{\partial t} I_{ij}(\mathbf{r}; \mathbf{r}', k_e, k_h) = -(\kappa_i + \kappa_j) I_{ij}(\mathbf{r}; \mathbf{r}', k_e, k_h)$$

$$- i\frac{\omega_0}{2k_0^2 \epsilon_r} (\Delta_\mathbf{r} - \Delta_{\mathbf{r}'}) I_{ij}(\mathbf{r}; \mathbf{r}', k_e, k_h)$$

$$-ig_0\frac{\sqrt{\nu_0}}{4\pi^2}\left(C_{ij}(\mathbf{r};\mathbf{r}',k_e,k_h)-C_{ji}^*(\mathbf{r}';\mathbf{r},k_e,k_h)\right.$$

$$+\frac{1}{k_0^2\epsilon_r}\left(\frac{\partial^2}{\partial r_j'r_k'}C_{ik}(\mathbf{r};\mathbf{r}',k_e,k_h)\right.$$

$$\left.\left.-\frac{\partial^2}{\partial r_ir_k}C_{jk}^*(\mathbf{r}';\mathbf{r},k_e,k_h)\right)\right) \tag{2.33}$$

with σ being the confinement factor along the z-direction,

$$\sigma=Q|\xi(r_z=z_0)|^2. \tag{2.34}$$

In Eq. (2.29) $\Lambda^{e,h}(\mathbf{r})$ describes carrier injection into the respective carrier states of the active semiconductor medium. $\gamma_{nr}f^e(\mathbf{r},k_e,k_h)$ is a rate for nonradiative recombination.

Using the equations presented above one can analyse the spatial and temporal coherence properties of a given quantum dot ensemble. This is of high importance for an investigation and fundamental analysis of the storage or transfer of quantum information relevant for e.g. quantum memories.

2.4 Quantum Theoretical Description

A full quantum theoretical treatment of a multi-level quantum dot ensemble coupled to light requires the simulation of the dynamics of the operators for the electromagnetic light fields, for the carriers and for the light fields. No further approximations as in the derivation of the quantum luminescence equations are done. The dynamics of the operators is then described by the Schrödinger equation with respect to a suitable Hamiltonian that includes all quantities and interaction processes relevant in the dynamics of the complex light–matter system (Buus and Flensberg, 2004; Ballentine, 1998). However, it is up to now out of reach to calculate the wave functions of the whole complex system represented by an active spatially inhomogeneous quantum dot system coupled to light. Instead, quantum-mechanical descriptions may consider reduced systems such as e.g.

the excitations of dots under the influence of Coulomb interactions (Hirose *et al.*, 2004), the interaction of a dot system in a near field optical system (Kobayashi and Ohtsu, 1999) or transport properties in quantum dots (Souza and Macedo, 2004). Other approaches concentrate on the density operator of a coupled atom-cavity system (Grond *et al.*, 2008) to analyse spin entanglements in quantum dots or express the density matrix in terms of the uncoupled atom and field states (Horak *et al.*, 1995) for an analysis of the stationary solution of a cascade laser. Coherent four-wave mixing in an ensemble of non-interacting atoms coupled to a radiation field has been described on the basis of a quantum-theoretical description using angular-momentum-like operators (Prior and Weiszmann, 1983).

In the quantum dot laser geometrical and system properties as well as material inhomogeneities play an important role in the dynamic interplay of light and matter. It is, however, out of reach to include these properties on a quantum-theoretical level. We will thus, in the following, restrict ourselves to the mesoscopic approach, representing a very reasonable compromise between exactness and computational effort.

References

Adachi, S., *J. Appl. Phys.* **58**, R1–R29, (1985).

Ballentine, L.E., *Quantum Mechanics: A Modern Development*, World Scientific Publishing, (1998).

Bester, G., Zunger, A., Wu, X. and Vanderbilt, D., *Phys. Rev. Lett.* **96**, 187602–187605, (2006).

Bilenca, A. and Eisenstein, G., *IEEE J. of Quantum Elect.* **40**, 690–702, (2004).

Bolcatto, P.G. and Proetto, C.R., *Phys. Rev. B* **59**, 12487–12498, (1999).

Boxberg, F. and Tulkki, J., *Rep. Prog. Phys.* **70**, 1425–1471, (2007).

Braskén, M., Lindberg, M., Sundholm, D. and Olsen, J., *Phys. Rev. B* **61**, 7652–7655, (2000).

Buus, H. and Flensberg, K., *Many-Body Quantum Theory in Condensed Matter Physics*, Oxford University Press, (2004).

Chow, W.W., Koch, S.W. and Sargent M., *III, Semiconductor-Laser Physics*, Springer-Verlag, Berlin, (1994).

Cornet, C., Schliwa, A., Even, J., Dore, F., Celebi, C., Letoublon, A., Mace, E., Paranthoen, C., Simon, A., Koenraad, P.M., Bertru, N., Bimberg, D. and Loualiche, S., *Phys. Rev. B* **74**, 35312–35320, (2006).

Deppe, D.G., Huffaker, D.L., Csutak, S., Zou, Z., Park, G. and Shchekin, O.B., *IEEE J. Quant. Electr.* **35**, 1238–1246, (1999).

Gehrig, E. and Hess, O., *Phys. Rev. A* **57**, 2150–2163, (1998).

Gehrig, E. and Hess, O., *Spatio-Temporal Dynamics and Quantum Fluctuations in Semiconductor Lasers.* Springer, Heidelberg, (2003).

Gehrig, E. and Hess, O., *Opt. Express* **16**, 3744–3752, (2008).

Grond, J., Pötz, W. and Imamoglu, A., **77**, 165307–165322, (2008).

Grosse, S., Sandmann, J.H., von Plessen, G., Feldmann, J., Lipsanen, H., Sopanen, M., Tulkki, J. and Ahopello, J., *Phys. Rev. B* **55**, 4473–4476, (1997).

Grundmann, M., Heitz, R., Bimberg, D., Sandmann, J.H.H. and Feldmann, J., *Phys. Status Solidi B* **203**, 121–132, (1997).

Grundmann, M., Stier, O. and Bimberg, D., *Phys. Rev. B* **52**, 11969–11981, (1995).

Grundmann, M. and Bimberg, D., *Phys. Rev. B* **55**, 9740–9745, (1997a).

Grundmann, M. and Bimberg, D., *Phys. Status Solidi A* **164**, 297–300, (1997b).

Hamm, J., Boehringer, K. and Hess, O., *Proc. of SPIE* **4646**, 176–189, (2002).

Hess, O., *Spatio-Temporal Dynamics of Semiconductor Lasers*, Wiss. - und Technik-Verlag, Berlin, (1993).

Hess, O. and Kuhn, T., *Prog. Quant. Electron.* **20**, 85–179, (1996).

Hirose, K., Meir, Y. and Wingreen, N.S., *Phys. E* **22**, 486–489, (2004).

Hofmann, H.F. and Hess, O., *Phys. Rev. A* **59**, 2342–2358, (1999).

Horak, P., Gheri, K.M. and Ritsch, H., *Phys. Rev. A* **51**, 3257–3266, (1995).

Huang, H. and Deppe, D.G., *IEEE J. Quantum Elect.* **37**, 691–698, (2001).

Huyet, G., O'Brien, D., Hegarty, S.P., McInerney, J.G., Uskov, A.V., Bimberg, D., Ribbat, C., Ustinov, V.M., Zhukov, A.E., Mikhrin, S.S., Kovsh, A.R., White, J.K., Hinzer, K. and SpringThorpe, A.J., *Phys. Status Solidi A* **201**, 345–352, (2004).

Jiang, H. and Singh, J., *J. Appl. Phys.* **85**, 7438–7442, (1999).

Kobayashi, K. and Ohtsu, M., *J. Microsc.* **194**, 249–254, (1999).

Mnkel, M., Kaiser, F. and Hess, O., *Phys. Lett. A* **222**, 67–75, (1996).

Oshiro, K., Akai, K. and Matsuura, M., *Phys. Rev. B* **59**, 10850–10855, (1999).

Prior, Y. and Weiszmann, A.N., *Phys. Rev. A* **29**, 2700–2708, (1983).

Qasaimeh, O., *IEEE Photonic Tech. L.* **16**, 542–544, (2004).

Souza, A.M.C. and Macedo, A.M.S., *Physica A* **344**, 677–684, (2004).

Stier, O., Grundmann, M. and Bimberg, D., *Phys. Rev. B* **59**, 5688–5701, (1999).

Sugawara, M., Mukai, K. and Shoji, H., *Appl. Phys. Lett.* **71**, 2791–2793, (1997).

Sugawara, M., Mukai, K., Nakata, Y., Ishikawa, H. and Sakamoto, A., *Phys. Rev. B* **61**, 7595–7603, (2000).

Sugawara, M., Akiyama, T., Hatori, N., Nakata, Y., Ebe, H. and Ishikawa, H., *Meas. Sci. Technol.* **13**, 1683–1691, (2002).

Tomic, S., Sunderland, A.G. and Bush, I.J., *J. Mater. Chem.* **16**, 1963–1972, (2006).

Uskov, A.V., Berg, W.T. and Mørk, J., *IEEE J. Quantum Elect.* **40**, 306–320, (2004).

Chapter 3

Light Meets Matter I: Microscopic Carrier Effects and Fundamental Light–Matter Interaction

The dynamic interplay of light and matter in active nano-structures is determined by a multitude of coupled physical processes in the active medium. In particular, the dynamic physical properties of an active nanomaterial are strongly related to the complex carrier dynamics in the charge carrier ensemble populating the dots. The charge carriers lead to a complex gain and index dynamic that affects the propagation and dynamic shaping of light. The optical fields, on the other hand, induce a complex excitation and relaxation dynamics within the charge carrier plasma. The physical processes involved in the mutual interplay of light and matter span a large regime ranging from femtoseconds (for the ultrashort carrier dynamics) up to the picoseconds (for the propagating light fields) and nanosecond (for the slow spatial carrier dynamics) regime. Thereby, the ultrafast carrier effects represent the microscopic seeds for the macroscopic behaviour of a given system.

With recent progress in both theoretical models and experimental femtosecond spectroscopy techniques, it is possible to investigate the nature of ultrafast carrier effects on a fundamental level. This allows one to identify and analyse the underlying physical interactions. Ultrafast scattering and spatio-spectral hole burning lead to continuous dynamic and nonlinear changes in the charge carrier plasma. In combination with geometrical boundaries (i.e. size and shape of the device) and the physical properties of the active medium they determine the complex and sometimes even chaotic light field dynamics and are thus responsible for the emission characteristics of a given device. It is thus highly desirable to understand the nature of these processes and, building on this knowledge, to design novel systems with improved properties.

In this chapter we will analyse the fundamental dynamic processes in the active charge carrier plasma. We will reveal the complex dynamics of the carriers in the dot levels (Section 3.1), investigate dynamic level hole burning (Section 3.2) and highlight the role of the nonlinear gain and index dynamics (Section 3.3) for the spatiotemporal light dynamics of propagating light fields. The simulations are based on the Maxwell–Bloch approach discussed in Chapter 2.

3.1 Dynamics in the Active Charge Carrier Plasma

Recent progress in the manufacturing of quantum dot (QD) materials with high quality and dot density has led to the experimental demonstration of quantum dot semiconductor optical amplifiers (QD-SOAs) with a high optical gain (Bakonyi *et al.*, 2003; Berg and Mørk, 2003) and a picosecond recovery of the saturated gain (Dommers *et al.*, 2007). Amplification of femtosecond pulses in a broad spectral range (Rafailov *et al.*, 2003) and pattern free amplification at Gbit rates (Akiyama *et al.*, 2003) using QD-SOAs have been achieved, showing their potential use in high speed optical networks.

These developments have pushed forward a large number of theoretical approaches aiming at a description and interpretation of ultrafast charge carrier effects in active nanomaterials. Quantum kinetic equations have been used to describe carrier–carrier and carrier–phonon mediated relaxation and dephasing in QD systems (Gartner *et al.*, 2006; Lorke *et al.*, 2006; Vu *et al.*, 2006). Analytical and numerical models based on rate equations have been used to study the amplification of ultrashort pulse-trains in QDSOAs (Berg and Mørk, 2004; Bilenca and Eisenstein, 2004; Qasaimeh, 2004) and to describe gain saturation and noise properties of QDSOAs

(Sugawara *et al.*, 2002; Uskov *et al.*, 2004). The Maxwell–Bloch approach (Gehrig and Hess, 2002) is particularly suited to explore ultrafast gain dynamics in QDSOAs (15) as well as carrier dynamics and spectral hole burning in quantum dot lasers and amplifiers (Chow and Koch, 2005; Gehrig and Hess, 2002).

The calculation of the carrier occupation probability of the QD states and the material polarisation within the frame of the quantum dot Bloch equations has been presented in Chapter 2. In this section, we will discuss in more detail the specific nature of the intra-dot and QD-WL scattering terms of the quantum dot Bloch equations. Thereby, the theoretical description includes the multi-level structure of bound dot states and the effect of carrier scattering between 2D wetting layer (WL) states and 0D QD states. Carrier–carrier and carrier–phonon scattering are included at the level of an effective rate approximation. We will, in particular, study the occupation probability of the QD levels and focus on the influence of phonon-mediated and Auger-type scattering processes.

Early measurements performed on QDs (e.g. electroluminescence and time-resolved photoluminescence) revealed relatively long carrier relaxation life-times comparable to the radiative recombination life-time (Mukai *et al.*, 1996). It leads to the assumption (termed 'phonon-bottleneck') that QD carrier relaxation is suppressed due to the discrete density of bound QD states and the narrow spectrum of LO-phonons. Subsequent experimental work has shown that QD carrier relaxation is efficient enough to allow the operation of optoelectronic devices via the ground state. Carrier scattering involving multiple phonons (Inoshita *et al.*, 1992; Magnusdottir *et al.*, 2002) as well as carrier–carrier scattering processes have been proposed as possible relaxation mechanisms (Magnusdottir *et al.*, 2003; Nielsen *et al.*, 2004; Uskow *et al.*, 1998).

3.1.1 *Intra-dot carrier scattering*

The scattering between confined dot carriers involves absorption and emission of LO phonons. The scattering rates are calculated according to (Li *et al.*, 1999) using the assumption that LO-phonons decay into acoustic phonons (due to an an-harmonic coupling term). The rates depend on the dephasing between the transition energy and the energy of the LO-phonon. (The frequency of LO-phonons in the InGaAs alloy is taken from (Bhattacharaya, 1993; Emura *et al.*, 1998)). The term describing intra-dot carrier–phonon scattering can be written as (Gehrig and Hess, 2002):

$$\frac{\partial}{\partial t} n_k^c \bigg|_{QD}^{c-ph} = \sum_{k_\uparrow > k} \{\gamma_{em}^{ph}(k, k_\uparrow) n_{k_\uparrow}^c (1 - n_k^c) - \gamma_{abs}^{ph}(k, k_\uparrow) n_k^c (1 - n_{k_\uparrow}^c)\}$$

$$- \sum_{k_\downarrow < k} \{\gamma_{em}^{ph}(k, k_\downarrow) n_k^c (1 - n_{k_\downarrow}^c)$$

$$- \gamma_{abs}^{ph}(k, k_\downarrow) n_{k_\downarrow}^c (1 - n_k^c)\}, \tag{3.1}$$

where c labels electrons or holes and k is the level index. $\gamma_{em}^{ph}(k, k_\uparrow)$ and $\gamma_{abs}^{ph}(k, k_\uparrow)$ are the scattering rates for the absorption and emission of one LO-phonon, respectively. n_k^c is the occupation probability of QD state k. The first bracket on the right side of Eq. (3.1) describes the relaxation of carriers from a higher level $k_\uparrow$ to a lower level k, the second bracket describes the excitation of carriers from a lower level $k_\downarrow$ to a higher level k.

3.1.2 *Phonon induced carrier scattering between quantum dots and wetting layer*

Another scattering process involving the emission or absorption of LO-phonons is the capture of carriers from the WL into the QDs or the escape of carriers from the QDs to the 2D wetting layer. The scattering rates for this type of process are modelled using the following equation (Chang *et al.*, 2004):

$$\frac{\partial}{\partial t} n_k^c \bigg|_{QD \leftrightarrow WL}^{c-ph} = \frac{m_c^\star}{4\pi\hbar^2} \frac{e^2 \omega_{LO}}{\epsilon_0} \left(\frac{1}{\epsilon_\infty} - \frac{1}{\epsilon_{stat}}\right)$$

$$\times [(n_{LO} + 1) f_Q(n_{2D}^c, E_Q)(1 - n_k^c)$$

$$- n_{LO}(1 - f_Q(n_{2D}^c, E_Q)) n_k^c] F(E_Q), \tag{3.2}$$

where $m_c^\star$ is the effective mass of the 2D carriers in the WL and ω_{LO} is the LO phonon frequency. ϵ_0 is the permittivity of vacuum, ϵ_∞ and ϵ_{stat} are the high frequency and static dielectric constants.

$n_{LO} = [\exp(\hbar\omega_{LO}/k_B T) - 1]^{-1}$ is the Bose occupation probability of LO phonons. $f_Q(n_{2D}^c, E_Q)$ is the Fermi occupation probability of the 2D carriers at the transition energy: $E_Q = E_{QD} + \hbar\omega_{LO}$. $F(E_Q)$ is a form-factor that depends on the transition energy and on the wave-functions of the 0D and 2D states. Generally, the scattering between the 2D wetting and the dot states depends on the respective transition energy (Reschner *et al.*, 2008).

3.1.3 *Auger scattering processes involving 0D and 2D carriers*

At high carrier densities, there is another relaxation processes that gains importance: Auger scattering involving the capture of one carrier in the dot. The corresponding capture rates of electrons and holes via Auger scattering (including carrier saturation effects) are modelled by the following equations (Uskow *et al.*, 1998):

$$\frac{\partial}{\partial t} n_i^e \bigg|_{QD\leftrightarrow WL}^{c-c} = C_{ee} n_{2D}^e n_{2D}^e (1 - n_i^e) + C_{eh} n_{2D}^e n_{2D}^h (1 - n_i^e)$$
$$+ \sum_j B_{eh} n_{2D}^h n_i^e (1 - n_j^h), \qquad (3.3)$$

$$\frac{\partial}{\partial t} n_j^h \bigg|_{QD\leftrightarrow WL}^{c-c} = C_{hh} n_{2D}^h n_{2D}^h (1 - n_j^h) + C_{he} n_{2D}^h n_{2D}^e (1 - n_j^h)$$
$$- \sum_i B_{eh} n_{2D}^h n_i^e (1 - n_j^h), \qquad (3.4)$$

where B_{he} is the Auger coefficient for a scattering event in which a 2D electron interacts with a 0D hole and is captured into the QD whereas the hole is scattered to the WL, C_{ee} is the Auger coefficient related to the interaction of two electrons from the WL and subsequent capture of one the electrons into the QD whereas the other is scattered into a 2D energy state of higher energy. The Auger coefficients are labelled according to the convention that the first index denotes the captured carrier and the second index denotes the carrier scattered to a 2D state.

Depending on the 2D carrier density in the WL and depending on the population of the dots with carriers the scattering may lead to carrier capture into the QDs or to ejection of carriers from the QDs. For high 2D charge carrier densities (of the order of 10^{11} cm^{-2}) the in-scattering of charge carriers dominates. Taking into account scattering processes between QD and WL discussed so far (see also Eqs. (3.2)–(3.4)), the dynamics of the WL carrier density is given by the following set of equations (Gehrig and Hess, 2002):

$$\frac{\partial}{\partial t} n_{2D}^e = J - D_a \left[\frac{\partial^2}{\partial x^2} + \frac{\partial^2}{\partial z^2} \right] n_{2D}^e - \Gamma_{loss}$$

$$- n_{QD} \sum_j \left\{ \frac{\partial}{\partial t} n_i^e \bigg|_{QD \leftrightarrow WL}^{e-ph} + \frac{\partial}{\partial t} n_j^e \bigg|_{QD \leftrightarrow WL}^{c-c} \right\},$$

$$(3.5)$$

$$\frac{\partial}{\partial t} n_{2D}^h = J - D_a \left[\frac{\partial^2}{\partial x^2} + \frac{\partial^2}{\partial z^2} \right] n_{2D}^h - \Gamma_{loss}$$

$$- n_{QD} \sum_j \left\{ \frac{\partial}{\partial t} n_j^h \bigg|_{QD \leftrightarrow WL}^{h-ph} + \frac{\partial}{\partial t} n_j^h \bigg|_{QD \leftrightarrow WL}^{c-c} \right\},$$

$$(3.6)$$

where the first term in both equations describes the carrier injection with current density J. D_a is the ambipolar diffusion coefficient and n_{QD} is the QD sheet density. The 2D carrier loss rate Γ_{loss} includes contributions due to non-radiative, spontaneous and Auger-recombination, respectively. It is given by: $\Gamma_{loss} = \gamma_{nr} n_{2D}^e + \gamma_{sp} n_{2D}^e n_{2D}^h + \gamma_{aug} n_{2D}^e n_{2D}^e n_{2D}^h$. The last two terms in Eqs. (3.5) and (3.6) describe scattering between carriers confined to QDs and 2D charge carriers.

For a detailed analysis of carrier relaxation dynamics it is necessary to include the carrier injection in a detailed and self-consistent way: for this purpose, we define the injection current density via the applied bias voltage V_{bias}. Assuming an Ohmic regime, we have: $J = \sigma E_\perp = \sigma V_{bias}/d_\perp$, where σ is the conductivity and $d_\perp$ is the device dimension perpendicular to the injection stripe. The conductivity is given by: $\sigma = e(\mu^e n_{3D}^e + \mu^h n_{3D}^h)$, where e is the electron charge, μ^e is the electron mobility, μ^h is the hole mobility, n_{3D}^e is the 3D electron charge density, and n_{3D}^h is the 3D hole charge density. Additionally, we approximate the 3D charge density by: $n_{3D} = n_{2D}/d_\perp$. Since we include radiative and non-radiative electron-hole recombination explicitly, we write the carrier injection in Eqs. (3.5) and (3.6) as:

$$J = e(\mu^e n_{2D}^e + \mu^h n_{2D}^h)\frac{V_{bias} - (\mu_{CB}^e - \mu_{VB}^h)}{d_\perp^2}, \qquad (3.7)$$

where $\mu_{CB}^e(n_{2D}^e)$ and $\mu_{CB}^h(n_{2D}^h)$ is the chemical potential of 2D electrons in the conduction band and 2D holes in the valence band, respectively.

3.1.4 *Level and gain dynamics*

Generally, the carrier scattering rates depend on the energy structure of the individual dots. As a consequence, the occupation probability of the quantum dot states in an inhomogeneously broadened QDSOA varies from dot to dot. As an example, Fig. 3.1 shows the occupation probability of the QD ground state for 100 'sample' dots from an inhomogeneously broadened device in a time-frame of 30 ps. The majority of the QDs shows a ground state population dynamics that is similar to a homogeneously broadened device, i.e. a rather uniform rise in occupation with increasing time. A subgroup of dots, however, shows a reduced rate of carrier in-scattering to the ground state. In these dots, the confinement energy of these QD with respect to 2D WL states exceeds the energy of one LO-phonon. As a consequence, carrier capture due to single LO-phonon scattering with 2D WL carriers is prohibited. In our sample quantum dot structure (see Fig. 3.1(c)) the energetic confinement of electron states is larger compared to that of hole states. This leads to a larger fraction of QDs with a reduced electron inscattering rate (compared to the hole in-scattering rate).

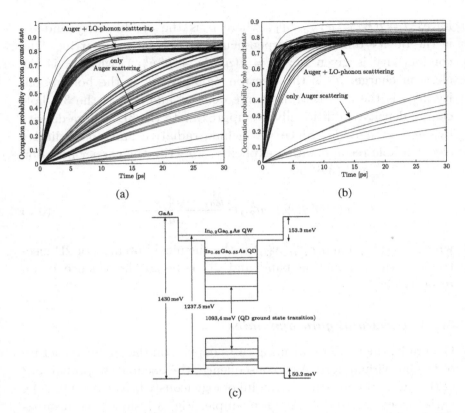

Fig. 3.1.   Occupation probability of the QD *electron ground state* (a) and QD *hole ground state* (b) for 100 'sample' QDs from the inhomogeneously broadened QDSOA. Relaxation via single phonon emission is prohibited for dots with carrier confinement energies exceeding the LO-phonon energy.

Having determined the energy level occupation probability of all QDs in the statistical ensemble the corresponding absorption/gain spectra can be calculated using the equation (Berg and Mørk, 2003):

$$g(\omega) = \frac{\pi}{2} \frac{k_0^2}{\beta} \frac{\Gamma}{\epsilon_0 \hbar} \frac{n_{QD} N_l}{d_a} \sum_{\kappa} \left[\sum_{i,j} \left(\Delta_{i,j}^{\kappa} \times \Delta_{i,j}^{\kappa *} \right) \mathbf{e_x} \right.$$

$$\left. \times \left[n_i^{e,\kappa} + n_j^{h,\kappa} - 1 \right] L(\omega, \omega_{j,i}^{0,\kappa}, \gamma_{j,i}^{\kappa}) \right] N_{sim}^{-1}. \qquad (3.8)$$

In Eq. (3.8) d_a is the width of the active area and N_l s the number of dot layers. The index κ runs over-all quantum dots in the statistical ensemble. i and j refer to the levels of electrons and holes, respectively. We assume s polarisation in the lateral (i.e. x-) direction. The dipole matrix element tensor thus is multiplied with the unit vector in x-direction $\mathbf{e_x}$. Furthermore, the following definitions have been used: the homogenous line-width (transition $i \leftrightarrow j$) of quantum dot κ is $\gamma_{j,i}^{\kappa} = \gamma_{j,i} + n_{2d}^e A_{j,i}^{e,\kappa} + n_{2d}^h A_{j,i}^{h,\kappa}$, the central frequency (transition $i \leftrightarrow j$) of quantum dot κ is $\omega_{j,i}^{0,\kappa} = \omega_{j,i}^{e,h,\kappa} + n_{2D}^e B_{j,i}^{e,\kappa} + n_{2D}^h B_{j,i}^{h,\kappa}$, and the normalised Lorentzian line function centred at $\omega_{j,i}^{0,\kappa}$, is given by $L(\omega, \omega_{j,i}^{0,\kappa}, \gamma_{j,i}^{\kappa}) = \gamma_{j,i}^{\kappa}(2\pi[(\omega - \omega_{j,i}^{0,\kappa})^2 + (\gamma_{j,i}^{\kappa})^2])^{-1}$.

The absorption and gain spectra of a homogeneously and inhomogeneously broadened QDSOA are summarised in Figs. 3.2(a) and 3.2(b), respectively. The first curve in the lower half-plane corresponds to the absorption spectra of empty QDs at 0 ps. The other curves show the absorption/gain spectra calculated at intervals of 0.75 ps. In the case of the ideal QDSOA (Fig. 3.2(a)) the ground state shows positive gain already after 3.0 ps. At the same time point the average gain due to the ground state transition of the inhomogenously broadened QDSOA is still negative (absorptive regime). This can be explained by the reduced carrier in-scattering rates of QD

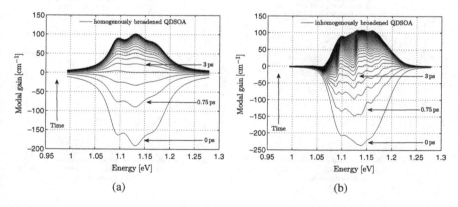

Fig. 3.2. Absorption/gain spectra of a homogeneously broadened QDSOA (a) and an inhomogenously broadened QDSOA (b) in a time-frame of 30 ps (the time interval between consecutive plots is 0.75 ps).

with higher confinement energy (that contribute mainly to the gain at the low energy end of the gain spectrum).

3.1.5 *Dynamics of carrier scattering rates*

In a next step we will discuss the dynamics of the intradot and dot-wetting layer scattering rates. For this purpose, we refer to an optically excited quantum dot ensemble and investigate the nature of the scattering dynamics within the charge carrier ensemble during relaxation. In our example we refer to the injection (at $t = 64$ ps) of an optical Gaussian shaped pulse ($\hbar\omega_{pulse} = 1.1$ eV) with a duration of 500 fs and an energy of 0.75 pJ. The ground state transition energy is 1.0934 eV.

The pulse induces a characteristic out-of-equilibrium situation in the charge carrier system. The resulting intra-dot relaxation and the carrier capture from the wetting layer then determines the re-population of the dot states with charge carriers.

Figure 3.3 visualises the *intra-dot* carrier-LO-phonon scattering rates calculated according to Eq. (3.1). Generally, the relaxation rates depend strongly on the difference between the LO-phonon energy and

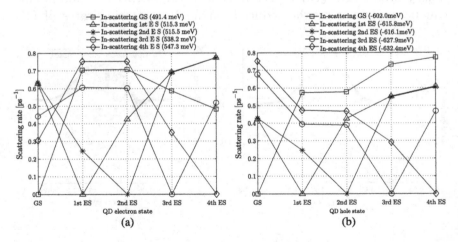

Fig. 3.3.   Intra-dot electron (a) and hole (b) in-scattering rate due to emission or absorption of one LO-phonon for each pair of initial and final states. The points are connected with guide lines.

the transition energy. As can be seen from Fig. 3.3(a) the effective electron relaxation can occur from the 1st and the 2nd excited state to the ground state (square marker). Therefore, the depletion of the ground state by the optical pulse is followed by a fast depletion of the 1st and 2nd excited electron states and 3rd and 4th excited hole states. These are precisely the QD levels that couple strongly to the ground state via LO-phonon scattering (see Fig. 3.3). States that couple only weakly to the ground state are depleted with a time delay (e.g. the 3rd excited electron state and the 1st and 2nd excited hole state).

In order to analyse the LO-phonon mediated refilling processes of the QD ground state after the passage of the probe pulse we can plot the total intra-dot capture rate for every electron and hole level. The results are summarised in Fig. 3.4. The solid line in Fig. 3.4(a) shows the capture rate related to the scattering of electrons from the excited QD states to the ground state. The negative capture rate (better termed emission rate) of the 4th excited state reveals that this state is in fact the main electron source for the ground state. Figure 3.4(b) shows that the hole ground state couples strongly to the 3rd and 4th excited hole states and therefore refills faster compared to the electron ground state.

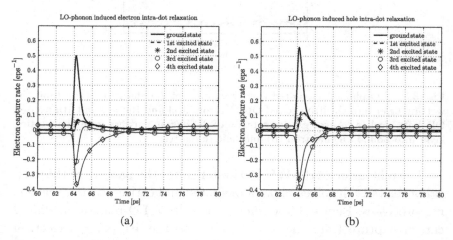

Fig. 3.4. LO-Phonon mediated intra-dot capture rates for each electron (a) and hole (b) level.

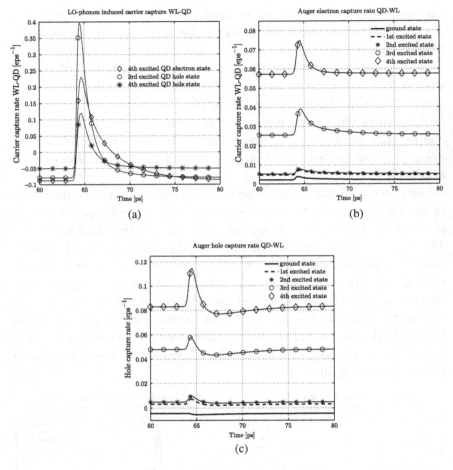

Fig. 3.5. LO-phonon mediated carrier capture of 2D charge carriers from the wetting layer into the 4th excited QD electron level and the 3rd and 4th excited QD hole level (a) Auger carrier capture of 2D charge carriers from the wetting layer into bound QD electron (b) and hole (c) states.

The relaxation of the WL-QD system after the passage to the optical probe pulse involves phonon mediated and carrier–carrier scattering processes. Figure 3.5(a) shows the rate of LO-phonon mediated carrier capture from the wetting layer to the 4th excited electron state and to the 3rd and 4th excited hole states. The rates are negative at steady state (before the injection of the optical pulse) indicating an out-scattering of carriers from the excited QD states to the wetting layer.

The pulse-induced carrier depletion of the QD states leads to a temporary increase of the carrier in-scattering from the wetting layer. This is also reflected in the dynamics of the Auger carrier capture rates from the wetting layer states into the dots (Figs. 3.5(b) and (c)). The Auger scattering between wetting layer states and 0D dot states thereby strongly depends on the confinement energy of the dots. For less confined carriers the scattering rates are higher compared to stronger confined carriers. At a 2D charge density of approximately 10^{12}cm^{-2}, Auger and LO-phonon mediated carrier capture rates involving 2D carriers in the wetting layer and 0D holes in exited QD states are of comparable magnitude.

3.2 Dynamic Level Hole Burning

The complex carrier scattering dynamics discussed so far strongly affects the dynamic occupation and depletion of the quantum dot levels. The level dynamics as calculated with the QD Maxwell–Bloch approach thus directly reflects the microscopic interactions occurring within the dots. As an example, we refer to a quantum dot semiconductor optical amplifier (length 2 mm, width 7 μm, emission wavelength 1260 nm for the ground state) consisting of 5 sheets of self-organised dots (InAs grown on GaAs, overgrown with a 5 nm InGaAs layer (Kovsh *et al.*, 2003)). The dot surface density in each layer is $4 \cdot 10^{10}$cm^{-2}. In the calculation we assume an inhomogeneous linewidth of 90 nm. The dots are assumed to have a pyramidal shape, with typical base length of 14 nm leading to three electron level energies (180, 100 and 85 meV, relative to the band edge of the conduction band in the wetting layer) and five hole level energies (110, 90, 75, 70 and 55 meV, relative to the band edge of the valence band in the wetting layer).

In the simulation the injection of the optical light pulse is realised via the boundary conditions of the light fields at the facets. Figure 3.6 visualises the dynamics of the carrier occupation in the three electron

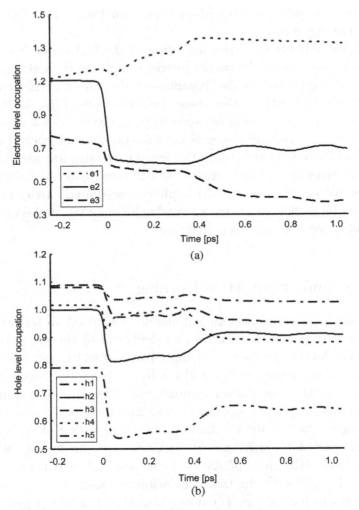

Fig. 3.6. Dynamics of level occupations in the electron (a) and hole (b) levels calculated at the centre of the output facet during the passage of the pulse.

levels (e1–e3) and five hole levels (h1–h5) during the passage of an ultrashort (150 fs) resonant light pulse. At the start of the calculation the injection current density has been set to $2 \cdot I_{thr}$. The input energy of the pulse is $E_{in} = 0.8 \cdot E_s$ where E_s is the pulse saturation energy (defined as the energy at which the gain has decreased

by 3 dB (Berg and Mørk, 2004)). The propagating resonant light
pulse induces highly non-equilibrium distributions in the charge carrier plasma and leads to a dynamic depletion of the levels. The specific shape of this dynamic level hole burning thereby depends on
injection current density, input pulse energy, dipole matrix elements
and intra-dot scattering in the multi-level structure. The microscopic
scattering processes involved in this 'level-burning' are determined by
emission and absorption of phonons, multi-phonon interactions and
the interaction with the carriers and phonons of the wetting layer discussed in the previous section. The magnitude of the various 'chan-
nels' for relaxation mechanisms thereby depend on the quantum dot
energy levels, on the energy difference to the surrounding layers and
on the coupling of a dot to its next neighbours. The dot-to-medium
and dot-to-dot interactions thereby are determined by the dot density and the light propagation that mesoscopically couples the dots.
In particular, the levels of a dot may 'talk' to each other via carrier
and phonon scattering. In the given example, the level separation
is near the LO phonon energy. As a result, a dynamic exchange of
carriers on the individual levels via emission and absorption of LO
phonons occurs. In particular, the refilling of the resonant dot levels from higher-energy levels (e.g. e3, h5) may lead to a partial gain
recovery. We note that this phonon interaction responsible for the
fast intra-dot redistribution of carriers is limited to level separations
approximately equal to the LO phonon energy: more deeply confined
dots with level separations much larger than this value would not display this behaviour but show decoupled level dynamics and selective
level hole burning instead. In order to visualise the dependence of the
influence of phonon interactions on operation conditions and material systems, Fig. 3.7 compares (for an input energy of $E_{in} = 0.6 \cdot E_s$)
the dynamics of the five holes for resonant (a) and off-resonant (b)
excitation. Figure 3.7(c) visualises the dynamics in a decoupled level
system (i.e. energy separation larger than the LO phonon energy).
Please note that we have (for a better visualisation) normalised the
level occupations to their initial values (i.e. before the passage of
the pulse). The strong influence of phonon emission and absorption processes lead to characteristic modulations in the coupled level

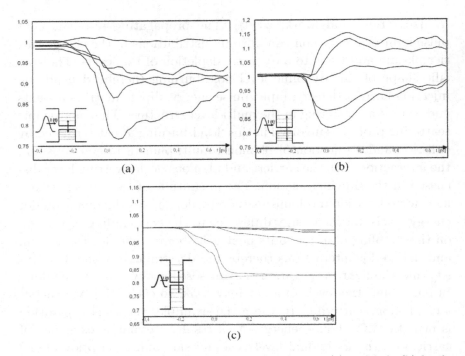

Fig. 3.7.   Dynamics of level occupations in the electron (a) and hole (b) levels calculated at the centre of the output facet during the passage of the pulse.

system (a), (b) where the level separation is near the LO phonon energy. Depending on the input energy, a dynamic depletion of all levels (a) or a selective depletion and filling of dot levels (b) occurs. In the case of the decoupled dot system a very selective depletion of individual levels (with energies near the pulse energy) takes place. The degree of coupling between the individual levels thus strongly effects the saturation behaviour of a QD laser waveguide that has been optically injected with a light pulse: for the same characteristics of an injected light pulse (i.e. input power, duration, pulse shape), a dot medium with decoupled level energies (i.e. energy separation larger than the LO phonon energy, will show a stronger saturation behaviour than a dot ensemble with a strong coupling of the level energies where a mutual exchange of level occupations via scattering and relaxation dynamics occurs (see also Chapter 6). These results clearly reveal the potential design of specific properties (e.g. design of the saturation degree of individual levels, small/strong influence of carrier–phonon scattering) of quantum dot semiconductor optical amplifier structures.

The ultrashort carrier dynamics is directly reflected in the microscopic gain, which can be derived from the (spatially averaged) microscopic distribution of the dipole density that is dynamically calculated within the framework of the QD Bloch equations. The computational results can then directly be compared to the results of an experimental pump–probe measurement. Figure 3.8 shows experimental (Gehrig *et al.*, 2004) and theoretical results on the dynamics of the gain for three different wavelengths of the injected light. The longest wavelength (1270 nm) corresponds to the amplification regime whereas the shortest wavelength (1216 nm) corresponds to the transparency regime. The intra-dot scattering (via emission and absorption of phonons) of the gain typically occurs on time scales of a few hundred fs up to a few ps leading to a fast partial recovery of the gain. Near transparency the partial exchange of inversion via absorption and emission of phonons leads to characteristic modulations in the gain dynamics that are directly correlated to the corresponding

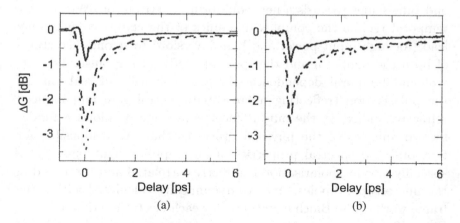

Fig. 3.8.   Calculated (a) and measured (b) (Rafailov *et al.*, 2003) gain change during propagation of a short pulse (150 fs) at 1216 nm (solid line), 1250 nm (dashed line) and 1270 nm (dotted line).

increased level dynamics within the charge carrier ensemble of the QDs. The modulation depth and period thereby depends on the energy separation of the levels as well as on injection current and input power. The temporal regime after approximately 1 ps is characterised by a comparatively slow re-establishment of the inversion via the injection current and the interaction with the carriers of the embedding layers. The calculation of the ultrashort time dynamics of the charge carriers clearly demonstrates that it is both operation conditions (e.g. injection current density and input power) and material properties (in particular the energetic separation of the energy levels) that affects the dynamic shaping as well as the saturation characteristics of a propagating light pulse.

3.3 Ultrashort Nonlinear Gain and Index Dynamics

A light pulse propagating in a semiconductor laser directly couples to the carrier populations in the energy levels. Thereby, it induces highly non-equilibrium distributions. As a consequence, dynamic changes in both gain and refractive index arise that relax on femto- and picosecond time scales towards a new equilibrium distribution. Within this temporal regime the complex dynamics of the charge carriers may lead (for high input power and high inversion) to strong distortions of both the temporal and the spectral profile of the pulse. The lateral and temporal dependence of the imaginary and real part of the polarisation (reflecting microscopic spectral gain and induced refractive index) at the output facet consequently allows a fundamental analysis of the physical processes that are responsible for temporal and spectral properties of an amplified light pulse. This spatially varying polarisation is directly correlated and composed of the microscopic dipoles that are dynamically calculated within the frame work of the Bloch equations (for each electron and hole level). Figure 3.9 summarises (in a time window of 1 ps) the dynamics of gain (negative imaginary part of the polarisation) and induced refractive index (negative real part of the polarisation) at the output facet

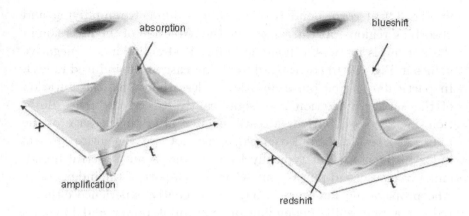

Fig. 3.9. Dynamics of imaginary (a) and real (b) part of the polarisation (calculated at the output facet of a quantum dot laser amplifier) during the propagation of an ultrashort (150 fs) pulse.

of a semiconductor laser amplifier that has been optically excited by an ultrashort light pulse (duration 150 fs). The bright spot on top of the figures indicates the corresponding dynamics (duration and position with respect to the displayed time axis) of the light pulse. In the calculation the injection current density has been set to 2.5 times the threshold current density, I_{thr}. The central pulse frequency is located within the amplifier gain bandwidth leading to an amplification of the pulse. During its propagation the pulse reduces the carrier distribution in the dots leading to dynamic level hole burning. The gain and induced refractive index are via the polarisation of the Bloch equation correlated to the dynamics of electrons and holes in the dots. They thus directly reflect the highly non-equilibrium carrier dynamics in the dots determined by hole burning, carrier injection as well as carrier–carrier and carrier–phonon scattering processes. The partial refilling of the hole via carrier relaxation and carrier injection typically occurs on time scales of a few hundred fs. These processes consequently determine the temporal and spectral shape of the 'spatio-spectral' trench burnt by the pulse in the spatio-spectral gain and index. The level hole burning and carrier heating induced by the pulse 'shape' the gain during the passage of the pulse (negative values in Fig. 3.9(a) correspond to high gain) leading to amplifying and absorbing regions. At the same time the reduction of the inversion in the dot levels induces a dynamic induced refractive index (negative values in Fig. 3.9(b) correspond to an increase in the induced index). In particular, if the pulse duration is less than 1 ps, the duration of the carrier relaxation may significantly exceed the pulse duration, i.e. the distributions are still excited after the passage of the pulse. As a result, the pulse envelope may not 'see' the entire spectral index dispersion (that typically leads to significant spectral broadening of picosecond pulses) but select only a part of the index curve. The propagating light pulse may consequently experience either a red- or a blue shift, depending on e.g. input power and injection current shaping the corresponding minima and maxima in the gain and index distributions. The spatio-temporally resolved calculation of the pulse-induced gain and index dynamics thus allows for a given laser geometry to directly obtain and analyse all informations on fundamental interaction processes that affect and shape the nonlinear amplification process and spectral dynamics of ultrashort pulses.

As an example we will in the following vary the width of the active area and analyse the resulting gain and index dynamics. Figure 3.10 shows for three values of the width of the active area the lateral and temporal dependence of the imaginary and real part of the polarisation reflecting gain and index at the output facet of a quantum dot laser amplifier during the propagation of an ultrashort laser pulse. Dark areas reflect high gain and index, whereas light shading reflect absorption and a reduction of carrier-induced index, respectively. The top row shows the corresponding intensity plot. Due to the nonlinear light–matter interactions, the pulse continuously 'shapes' its gain and index distribution during propagation. This leads to characteristic temporal modulations in both gain and index. Since carrier relaxation occurs on a time scale of a few hundred fs, the dynamic gain and index distribution are still excited after the passage of the pulse. The temporal shift in gain and index visualise the pulse-induced gain (dark shading) within the pulse as well as induced index dispersion resulting from the dynamics of the real and imaginary part of the inter-level polarisation. In the quantum dot laser amplifier

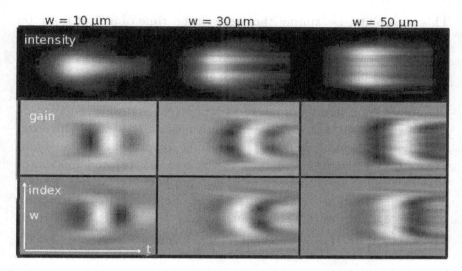

Fig. 3.10. Dynamics of intensity (*top row*), imaginary (*middle row*) and real part (*bottom row*) of the polarisation during the propagation of a 150 fs pulse in dependence on stripe width.

with cavity width w $= 10\,\mu$m the distributions are rather uniform. As a consequence the pulse will be characterised by a uniform lateral beam profile. An increase in lateral width leads to characteristic modulation in transverse direction. In particular, these structures show a characteristic curvature in lateral dimension. This is a direct consequence of the intensity dependent carrier depletion in the levels induced by the spatial profile of the pulse. It is these optical pattern of dynamic gain and index that shape the pulse in the next time step. In particular, Fig. 3.10 shows that the distributions within the quantum dot device are quite uniform in a lateral direction up to a stripe width of $10\,\mu$m, whereas a corresponding quantum well device of the same stripe width would exhibit lateral modulations responsible for filamentation. This is an important result for the interpretation of beam quality that will be discussed in Chapter 5.

Additionally, the dynamic interplay of gain saturation and index dispersion directly affects both amplitude and phase of a propagating signal pulse. Depending on the excitation conditions this may thus even lead to a speed up or slow down of the propagating light pulse.

This possibility, i.e. tuning the propagation time of a light pulse by modifying the nonlinear index and gain has recently attracted attention as it opens ways to the development of new optimal components with optical memory capability (see also Chapter 6).

In the following we will have a closer look at the dependence of the polarisation dynamics on injection current and pulse energy.

We will consider the following two situations: (i) strongly inverted amplifying medium and (ii) absorbing quantum dot medium. Figure 3.11 shows the dynamics of the real (a), (c) and imaginary part (b), (d) of the polarisation in the amplification regime. Shown are the distributions at the output facet of the amplifier during the passage of a light pulse. The input power is $P_{in} = 0.1\,P_s$ (a), (b) and $P_{in} = 1\,P_s$ (c), (d). Negative values in the real part, P', correspond

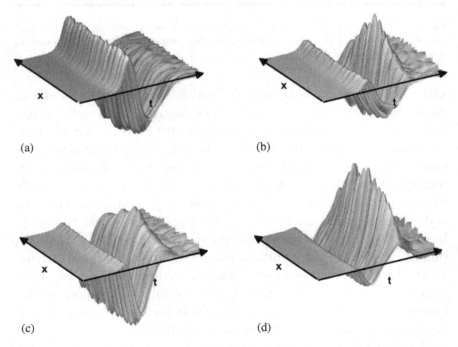

Fig. 3.11.   Dynamics of the real part (a), (c) and the imaginary part (b), (d) of the polarisation in the amplification regime. The distributions were calculated at the output facet of the quantum dot laser for an input power of $P_{in} = 0.1\,P_s$ (a), (b) and $P_{in} = 1\,P_s$ (c), (d).

to an increase in refractive index whereas positive values refer to a reduction in the carrier-induced refractive index. P' shows the typical dispersion shape. The spatial shape thereby directly reflects the intradot dynamics: in the case of high input power levels, the resonant light pulse induces a strong level hole burning in the ground state levels. This depletion then causes a partial refilling via intradot scattering (from higher energy states) and carrier capture from the wetting layer states. This procedure generates a characteristic spectral dispersion that is via the propagation of the light fields directly transferred to corresponding spatial profiles. In the given example this leads to a characteristic rise of the spatial index in the front part of the pulse and to an index reduction near the trailing edge of the pulse (Fig. 3.11(c)). The imaginary part (P'') reflects the pulse-induced gain shaping (negative values correspond to high gain). For high input powers (Fig. 3.11(d)) the amplification regime (i.e. negative values) shifts to earlier times and a pronounced absorbing region arises near the trailing pulse part. This directly reflects the complex intra-dot dynamics: The front part of the pulse induces stimulated emission processes in the charge carrier system. Due to the finite time scales of the carrier capture from higher levels and wetting layer states the trailing part of the pulse sees the strongly depleted levels and consequently is partially absorbed (positive values in the gain). Input power levels that are sufficiently high to saturate the amplifier thus cause strong distortions in the index dispersion and gain dynamics (Fig. 3.11(c,d)). This may eventually lead to a dynamic acceleration of the propagating pulse (see Chapter 6). The situation is significantly changed when the amplifier is driven in the absorption regime (Fig. 3.12). In this case, the injected light pulse leads to a carrier accumulation and the index dispersion changes its sign (Fig. 3.12(a,c)). Near saturation (Fig. 3.12(c)), the index dispersion and the strong absorption in the leading part of the pulse can eventually lead to a slow-down of the light pulse.

The observed effects can be summarised as follows:

(i) The complex carrier dynamics that is typical for spatially inhomogeneous quantum dot ensemble leads to dynamic changes

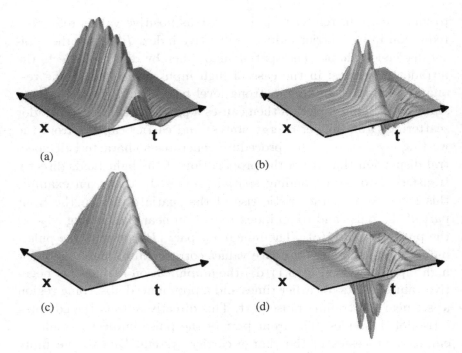

Fig. 3.12. Dynamics of the real part (a), (c) and the imaginary part (b), (d) of the polarisation in the absorption regime: The distributions were calculated at the output facet of the quantum dot laser for an input power of $P_{in} = 0.1\,P_s$ (a), (b) and $P_{in} = 1\,P_s$ (c), (d).

in gain and index that do not necessarily display the typical resonance-like behaviour but generally show a complex gain dynamics and distorted index dispersion. This originates from the difference in the time scales on which the underlying physical processes take place. These are, in particular, ultrafast (fs...ps) carrier and carrier–phonon scattering processes within the dots, propagation of the light pulse (ps regime) and slow re-establishment of the spatial inversion due to the injection current (ns regime).

(ii) The observed gain and index dynamics is dynamically generated by the spatially dependent light–matter interactions occurring during the propagation of the pulse.

(iii) The spatiotemporally varying gain and index reveal the reasons for changes in the propagation times: reshaping of the gain (i.e. saturation) induces a shift of the pulse centre. Reshaping of the index leads to velocity changes and to the generation of new frequencies within the pulse shape. Both effects affect the propagation time and depend on current and power.

The influence of the complex interplay of index dispersion and gain on relevant output quantities such as passage time (dynamic acceleration and slow-down) and dynamic shaping of the propagating light pulse will be discussed in Chapter 6.

3.4 Conclusion

We have analysed and discussed the ultrafast carrier dynamics and nonlinear light–matter coupling in active quantum dot laser waveguides. Simulation results on the basis of Maxwell–Bloch equations visualised the ultrafast carrier dynamics and dynamic light–matter coupling, affecting the optical properties of the nano-devices. They reveal nonlinear effects such as, in particular, carrier scattering, spatio-spectral saturation and level hole burning.

References

Akiyama, T., Hatori, N., Nakata, Y., Ebe, H. and Sugawara, M., *Phys. Status Solidi B* **238**, 301–334, (2003).

Bakonyi, Z., Su, H., Onishchukov, G., Lester, L.F., Gray, A.L., Newell, T.C. and Tuennermann, A., *IEEE J. Quantum Elect.* **39**, 1409–1413, (2003).

Berg, T.W. and Mørk, J., *Appl. Phy. Lett.* **82**, 3083–3086, (2003).

Berg, T.W. and Mørk, J., *IEEE J. Quantum Elect.* **40**, 1527–1539, (2004).

Bhattacharaya, P., (Ed.), *Properties of Lattice-Matched and Strained Indium Gallium Arsenide*, Inspec, UK, (1993).

Bilenca, A. and Eisenstein, G., *IEEE J. Quantum Elect.* **40**, 690–702, (2004).

Chang, S., Chuang, S. and Holonyak, N., *Phys. Rev. B* **70**, 1–12, (2004).

Chow, W.W. and Koch, S.W., *IEEE J. Quantum Elect.* **41**, 495–505, (2005).

Dommers, S., Temnov, V.V., Woggon, U., Gomis, J., Martinez-Pastor, J., Laemmlin, M. and Bimberg, *Appl. Phys. Lett.* **90**, 33508–33511, (2007).

Emura, S., Gonda, S., Matsui, Y. and Hayashi, H., *Phys. Rev. B* **38**, 3280–3286, (1988).

Gartner, P., Seebeck, J. and Jahnke, F., *Phys. Rev. B* **73**, 115307, (2006).

Gehrig, E. and Hess, O., *Phys. Rev. A* **65**, 33804–33819, (2002).

Gehrig, E., Poel, M.V.D., Birkedal, D., Hvam, J. and Hess, O., *ECOC 2004 Proc.* **3**, 616, Stockholm, Sweden.

Haug, H. and Koch, S.W., *Quantum Theory of the Optical and Electronic Properties of Semiconductors*, World Scientific, Singapore, (1998).

Inoshita, T. and Sakaki, H., *Phys. Rev. B* **46**, 7260–7263, (1992).

Kovsh, A.R., Maleev, N.A., Zhukov, A.E., Mikhrin, S.S., Vasil'ev, A.P., Semenova, E.A., Shernyakov, Y.M., Maximov, M.V., Livshits, D.A., Ustinov, V.M., Ledentsov, N.N., Bimberg, D. and Alferov, Z.I., *J. Crystal Growth* **250**, 729–736, (2003).

Li, X., Arakawa, Y. and Nakayama, H., *Phys. Rev. B* **109**, 351–356, (1999).

Lorke, M., Nielsen, T.R., Seebeck, J., Gartner, P. and Jahnke, F., *Phys. Rev. B* **73**, 85324–85333, (2006).

Magnusdottir, I., Uskov, A.V., Bischoff, S. and Mørk, J., *J. Appl. Phys.* **92**, 5982–5990, (2002).

Magnusdottir, I., Bischoff, S., Uskov, A.V. and Mørk, J., *Phys. Rev. B* **67**, 205326–205330, (2003).

Mukai, K., Ohtsuka, N., Shoji, H. and Sugawara, M., *Phys. Rev. B* **54**, 5243–5246, (1996).

Nielsen, T.R., Gartner, P. and Jahnke, F., *Phys. Rev. B* **69**, 1–13, (2004).

Qasaimeh, O., *IEEE Photonics Tech. Lett.* **16**, 542–544, (2004).

Rafailov, E.U., Loza-Alvarez, P., Sibbett, W., Sokolovskii, G.S., Livshits, D.A., Zhukov, A.E. and Ustinov, V.M., *Photonics Tech. Lett.* **15**, 1023–1025, (2003).

Reschner, D.W., Gehrig, E. and Hess, O., in *IEEE J. Quantum. Elect.* **45**, 21–33, (2009).

Sugawara, M., Akiyama, T., Hatori, N., Nakata, Y., Ebe, H. and Ishikawa, H., *Meas. Sci. Techno.* **13**, 1683–1691, (2002).

Uskow, A.V., J. McInerney, F. Adler, H. Schweizer and M.H. Pilkuhn, *Applied Physics Letters*, **72**, 58–60, (1998).

Uskov, A.V., Berg, T.W. and Mørk, J., *IEEE J. Quantum Elect.* **40**, 306–320, (2004).

van der Poel, M., Gehrig, E., Hess, O., Birkedal, D. and Hvam, J.M., *IEEE J. Quantum Elect.* **41**, 1115–1123, (2005).

Vu, Q.T., Haug, H. and Koch, S.W., *Phys. Rev. B* **73**, 205317–205325, (2006).

Chapter 4

Light Meets Matter II: Mesoscopic Space-Time Dynamics

4.1 Introduction: Transverse and Longitudinal Mode Dynamics

In spatially extended active nano-structures the dynamic interplay of spatial with temporal degrees of freedom may lead to diverse and complex dynamics. The physical processes that are responsible for this dynamic nonlinear coupling are, in particular, light diffraction, carrier diffusion, scattering and (radiative and nonradiative) recombination. The resulting nonlinear light–matter coupling leads to characteristic light field and carrier distributions with imprinted variations in both space and time.

In addition to these physical processes occurring during the interaction of light and matter, the geometrical constraints of a device or waveguide structure have, by enhancing some modes and suppressing others, a strong influence on the resulting dynamics of the system. However, not only the magnitude of the *transverse* extension of a device (such as a laser or waveguide) but also the nanostructure of quantum dot materials affects to a high degree the *transverse mode dynamics* that appears in the formation and migration of dynamic filaments and the (spatial and spectral) beam characteristics (nearfield, farfield, optical spectra). Generally, the length of the laser device determines the number and interplay of longitudinal modes and spatio-spectral hole burning. The *longitudinal* propagation dynamics

71

thus plays an important role in the spectral dynamics and the modulation properties of a laser (i.e. applying a temporally varying injection current). Again, via the nonlinear coupling to the active quantum dot nanomaterial, the transverse and longitudinal degree of freedom influence each other. The dynamics as well as the emission spectra as observed in experiments are, indeed, the result of a complex interplay of transverse modes and the light fields propagating in longitudinal direction.

Such spatio-temporal effects are thus important for both edge- and vertical-cavity surface-emitting laser (VCSEL) semiconductor laser structures. In this chapter we will focus our discourse on edge-emitting lasers, while the properties of surface emitting structures will be discussed in Chapter 9.

4.2 Influence of the Transverse Degree of Freedom and Nano-Structuring on Nearfield Dynamics and Spectra

Semiconductor lasers with a large active area (e.g. 50–150 μm) are key to a generation of high output power. However, as their lateral width significantly exceeds the laser-internal interaction lengths (typically lying in the μm-regime) given by light diffraction and carrier diffusion the spatio-temporal behaviour of the nearfield displays a complex transverse light field dynamics.

In the following we will discuss selective results in order to particularly visualise the dynamic interplay of light and matter within the active area. Our simulations are based on the mesoscopic quantum dot Maxwell–Bloch equations and include the full coupled spatio-temporal light field and inter/intra-level carrier dynamics. Intradot scattering via emission and absorption of phonons, as well as the scattering with the carriers and phonons of the surrounding wetting layer are dynamically included on a mesoscopic level. The time-dependent calculation of the carrier and light field dynamics allow for explicit consideration of the individual time scales of the various interaction processes. The relevant time scales range from the femtosecond

(a) w = 10 μm

(b) w = 30 μm

(c) w = 50 μm

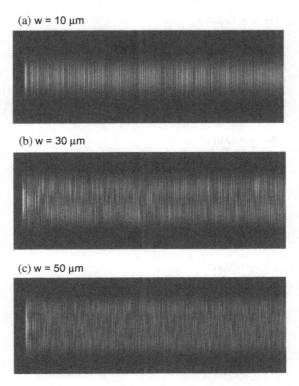

Fig. 4.1.   From single-stripe to broad-area geometry: influence of the transverse degree of freedom on the spatio-temporal dynamics in quantum dot lasers. The stripe width is (a) 10 μm, (b) 30 μm and (c) 50 μm. Bright shading indicates high intensity values.

regime (for the fast carrier scattering processes) up to the picosecond and nanosecond regime (for the dynamics of the propagating light fields and of the spatial carrier density).

Figures 4.1 and 4.2 display (for a time window of 10 ns) the temporal behaviour of the light field distribution at the output facet of a quantum dot (Fig. 4.1) and a quantum well laser (Fig. 4.2), respectively. The material and waveguide design are identical in both systems. Shown are results for a stripe width of (a) 10 μm, (b) 30 μm and (c) 50 μm. The cavity length is 1.3 mm. Bright shading represents high intensity values. In the simulations the same material parameter

(a) w = 10 µm

(b) w = 30 µm

(c) w = 50 µm

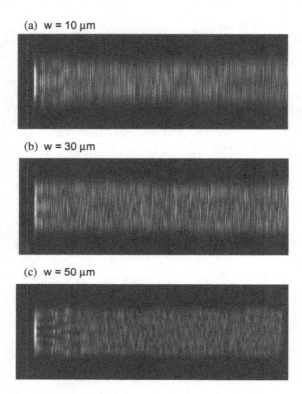

Fig. 4.2.   From single-stripe to broad-area geometry: influence of the transverse degree of freedom on the spatio-temporal dynamics in quantum well lasers. The stripe width is (a) $10\,\mu$m, (b) $30\,\mu$m and (c) $50\,\mu$m. Bright shading indicates high intensity values.

set is used for both systems. For the quantum dot laser a dot density of $10^{11}\,\mathrm{cm}^{-2}$ has been assumed. The figures clearly reveal the influence of both the geometry and structuring of the medium.

For a stripe width of $10\,\mu$m the quantum dot laser (Fig. 4.1(a)) shows a laterally uniform nearfield dynamics. For the same stripe width, the quantum well laser (Fig. 4.2(a)) is characterised by a dynamic interplay and coexistence of higher order modes leading to a reduced spatio-spectral purity. In both systems, switching of the injection current from zero to a multiple of the threshold value initiates characteristic relaxation oscillations in the emitted intensity that can be seen in the temporal direction. After some nanoseconds the laser reaches a quasi-stationary state. However, the interplay of the various time scales of light propagation, carrier scattering and diffusion will always imply the existence of small fluctuations in the light intensity.

Generally, a reduction of the lateral extension of the active stripe leads to a reduced influence of the transverse degree of freedom. The narrow stripe width prevents filamentation and higher order transverse modes cannot evolve. In quantum well lasers, a typical transverse dimension for single-mode operation is $5\,\mu$m. The laser stripe then supports only one transverse mode leading to a Gaussian shaped beam-profile. In quantum dot lasers, the strong localisation of the inversion in the dots and the low amplitude phase coupling (alpha factor) leads to a further reduction of the influence of the lateral degree of freedom and thus to a strong reduction of the transverse light field dynamics. As a result, the transition to multi-mode emission occurs at larger stripe widths (typically: $10\,\mu$m). This can be seen in Fig. 4.1(a) and Fig. 4.2(a) where the stripe width has been set to $10\,\mu$m.

An enlargement in stripe width leads in both systems to an increase in output power but at the same time to an increase in the dynamic interplay of light diffraction, carrier diffusion and scattering. The heightened complexity of the system eventually leads to characteristic transverse migration of the light fields and to filamentation. As Fig. 4.1(a) and Fig. 4.2 show, the influence of the transverse degree of freedom thereby is significantly higher in a quantum well system than in a quantum dot laser.

The difference in the influence of the transverse degree of freedom becomes particular pronounced if we compare the emission of quantum dot and quantum well devices with larger stripe widths. As an example, Figs. 4.1 and 4.2 (b)–(c) exhibit the dynamics of the intensity at the output facet of the lasers with a stripe width of $30\,\mu$m (b) and $50\,\mu$m (c), respectively. In spite of the identical pump current density, the light fields dynamics in the quantum well system (Fig. 4.2) is characterised by a much higher complexity. In particular, the figures visualise a characteristic zig-zag movement occurring on picosecond time scales. This behaviour originates from the transverse degree of freedom which gains importance for larger stripe widths: the dynamic interplay of light diffraction and carrier diffusion results in a complex transverse light field dynamics and to the migration of optical filaments (in the μm regime).

A comparison of Figs. 4.1 and 4.2 shows that a nano-structuring of the active medium leads to a significant reduction in the complexity of the light field dynamics: the strong carrier localisation in the dots and the small amplitude-phase coupling (see Chapter 5) lead to a strong reduction of the influence of the transverse extension of the device. As a consequence, nano-structured active devices with a stripe width of 10 μm still show a single-mode emission.

The complex light field dynamics that can be seen in Figs. 4.1 and 4.2 is the origin of characteristic intensity modulations that can be observed in experimental investigations (Adachihara *et al.*, 1993; Burkhard *et al.*, 1999; Gehrig and Hess, 2000; Marciante and Agrawal, 1997). Furthermore, it is responsible for the spatial and spectral purity of the emitted radiation (see Chapter 5). We thus can conclude that a nano-structuring of the active semiconductor medium allows for a control and improvement of both spatial and spectral beam quality.

In the next step we will discuss the emission properties of active semiconductor devices with different injection level. Generally, the injection of carriers into the active area leads to a rise in both spatial carrier transfer and local gain and index changes. Furthermore, the high inversion provides sufficient gain to support a larger number of modes. Increasing the injection current density thus leads to an increase in mode dynamics. As an example, Fig. 4.3 shows results of a calculation of spatially resolved emission spectra as obtained from a Fourier transformation of the emitted light fields at the output facet. The length of the cavity was $L = 3$ mm, the width was $w = 30$ μm. The figures refer to an injection level of approximately $1 \times$ the threshold current density (a) and $3 \times$ the threshold current density (b). The first row refers to the quantum dot system and the second row shows spectra of a corresponding quantum well system.

The quantum well system is characterised by a number (in the example: three) of longitudinal modes whose separation depends on

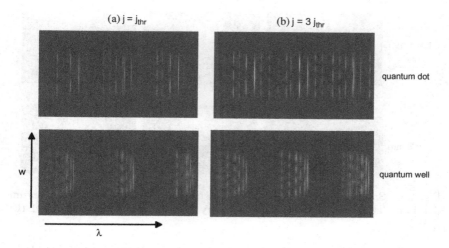

Fig. 4.3. Emission spectra: influence of the current density on the spectral properties of quantum dot (*top*) and quantum well (*bottom*) lasers. The stripe width of both lasers is $30\,\mu$m.

the cavity length. Each mode is composed of a group of transverse modes. The modes of higher transverse order can be found in direction of higher energy. In the quantum dot laser, fewer modes exist for the same current density. Please note that in the case of a quantum dot system a particular transverse mode (e.g. with three maximum in lateral direction) may occur more that once within a mode group. This originates from the fact that more than one transition (with individual transition energy) may be possible in a given quantum dot structure. The computational results demonstrate an increase in mode number with increasing carrier injection level. Furthermore, they highlight, once more, that a nano-structuring of the active area leads to a reduction in the number of coexisting modes.

4.3 Longitudinal Modes

In order to illustrate the influence of the longitudinal degree of freedom we will in the following consider snapshots ($\Delta t = 25\,$ps) of the intensity distribution in the active area of a quantum dot laser ($w = 10\,\mu$m) with a cavity length of $L = 1\,$mm (a) and $L = 2\,$mm (b),

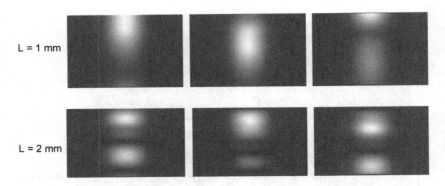

Fig. 4.4.   Snapshots ($\Delta t = 25\,\text{ps}$) of the intensity in the active area of a quantum dot laser (stripe width $10\,\mu\text{m}$) for a cavity length of (a) $L = 1\,\text{mm}$ and (b) $L = 2\,\text{mm}$.

respectively. Generally, the dynamic coupling of light and matter lead to the formation of dynamic optical patterns with typical extensions of several tens or even hundreds of μm. Depending on the propagation length, one or more than one 'light bullet' may exist for a given time step. The probability for the coexistence of several light spots thereby increases with cavity length. The dynamic hole burning induced by the light fields thereby leads to regions of higher gain surrounding the spatial hole. As a result, a second mode may evolve. We would like to note, however, that the number of the 'light bullets' in a given time step does not have to be identical to the number of coexisting longitudinal modes (see also Fig. 4.3). The time-integrated emission spectra includes all modes that principally exist within the considered time interval (in our example: $50\,\text{ns}$). However, not all of the modes that can be seen in the spectra do exist at every time step. It is much more likely that modes may vanish from time to time and then re-appear. This is an important aspect in the dynamic interplay and gain competition of modes.

4.4 Coupled Space-Time Dynamics in the Active Area

Typically, the spatio-temporal light field and carrier dynamics play a major role for relevant physical quantities such as the spatio-spectral gain and induced index of the system that, in combination with the complex carrier dynamics, determine output quantities of the laser system (relevant output properties are, e.g. emission wavelength, spectral bandwidth, saturation properties and temporal emission characteristics). The space-time simulation gives us the privilege to take a look inside a nano-structured laser by visualising the laser-internal light field dynamics. In the following, we will illustrate the influence of injection level and geometry (4.4.1), discuss the role of disorder (4.4.2) and present theoretical extensions required for a fundamental analysis of light fluctuations (4.4.3). Without loss of generality we refer to an InGaAs quantum dot laser with a density of quantum dots of $10^{10}\,\mathrm{cm}^{-2}$.

4.4.1 *Influence of injection level and geometry*

We show numerical results on the spatio-temporal dynamics in quantum dot lasers with a lateral extension of the active area of $10\,\mu\mathrm{m}$ (Fig. 4.5), $30\,\mu\mathrm{m}$ (Fig. 4.6) and $50\,\mu\mathrm{m}$ (Fig. 4.7), respectively. The cavity length is $2\,\mathrm{mm}$. In a first step, we investigate the dynamics in a quantum dot laser with a width of $10\,\mu\mathrm{m}$. The discussion in the previous section has shown that the transverse dynamics in this system is negligible if the device is driven in laser operation

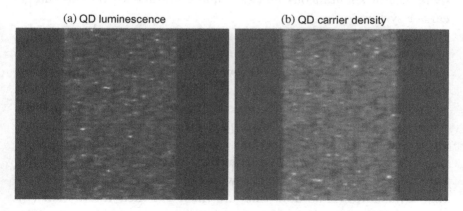

Fig. 4.5. Snapshots of (a) luminescence and (b) carrier density in a quantum dot laser driven below threshold ($w = 10\,\mu\mathrm{m}$).

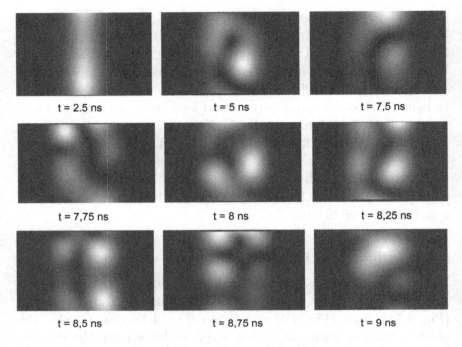

t = 2.5 ns	t = 5 ns	t = 7,5 ns
t = 7,75 ns	t = 8 ns	t = 8,25 ns
t = 8,5 ns	t = 8,75 ns	t = 9 ns

Fig. 4.6.   Light field dynamics in a quantum dot laser ($w = 30\,\mu$m).

(i.e. above threshold). We thus use this system to visualise the dynamics of spontaneous emission and its influence on the charge carrier system.

Figure 4.5 shows (a) the intensity and (b) the carrier density (sum of electrons and holes) for a quantum dot ensemble where the initial filling of the dot levels is below the threshold occupation necessary for inversion. The initial carrier injection via the contact leads to spontaneous and induced emission processes. In this calculation we have assumed the very ideal case where each dot has the same electronic structure (i.e. identical dot size, level energies and dipole matrix elements) and where the distribution of the dots in the structure is uniform. In our example we assume a pumping of the wetting layer states. The dots thus are filled via their interaction (by dynamic inscattering processes) with the wetting layer states. The formation of characteristic optical patterns is a direct consequence

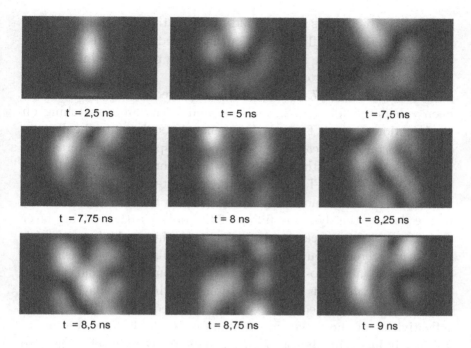

Fig. 4.7.   Light field dynamics in a quantum dot laser ($w = 50\,\mu$m).

of spontaneous light fluctuations and scattering. The microscopic intra-dot scattering of the carriers within the dots via emission and absorption of phonons, the interaction of the 'dot carriers' with the carriers and the phonons of the wetting layer and the nonlinear coupling to the propagating light fields lead to a spatially varying occupation of the dots and subsequently to complex transverse carrier and light field dynamics. It is important to note that the interplay of light with the carriers results in a spatio-temporally varying occupation although we have assumed the 'ideal' case of uniform carrier injection and regular matrix-like positioning of identical quantum dots (i.e. of equal size, level energies, matrix elements).

As a next example we will consider the dynamics in a quantum dot ensemble (width $30\,\mu$m and $50\,\mu$m) where the dots are initially excited well above their transparency level (Figs. 4.6 and 4.7). These figures show the influence of a high carrier injection (above threshold) which leads to the build-up of spatio-temporal coherence and reduces the local fluctuation in carrier and light fields that exist near threshold. At the same time, the figures reveal the influence of the transverse degree of freedom.

In Figs. 4.6 and 4.7, the snapshots displayed in the first row have been calculated for $t = 2.5\,\text{ns}$, $5\,\text{ns}$, and $7.5\,\text{ns}$ after switching on the injection current. The remaining snapshots illustrate the dynamics on shorter time scales ($\Delta t = 25\,\text{ps}$). In spite of the increase in transverse stripe width the intensity is rather uniform when compared to Fig. 4.5(a). This uniformity originates from induced emission processes which now play a major role in the over-all behaviour of the device: the dynamic filling of the dots via the wetting layer states establishes a carrier inversion and thus a high gain leading to an increased influence of induced emission processes. As a consequence, a spatio-temporal coherence builds up that via the propagating light fields is transferred in both time and spatial dimensions. The strong injection level and the feedback realised by the reflectivities of the facets in combination with the dynamic waveguiding induced by the propagating light fields themselves lead to the built-up a spatio-temporal coherence. As a result, the dynamic transversely and longitudinally varying coherent light patterns show a characteristic dynamics and migration in time but do not show the sub-structure that can be seen in the system near threshold (Fig. 4.5).

The snapshots calculated for the first time steps show the start-up behaviour: light from spontaneous and induced emission propagates in the resonator. After some round trips the interaction of the counterpropagating light fields with the active charge carrier plasma lead to the formation of coherent 'light bullets'. On longer time scales (second and third row of Figs. 4.6 and 4.7), i.e. after the first relaxation oscillations, a complex transverse light field dynamics begins. The dynamic light–matter coupling and the mutual interplay of self-focusing, light diffraction and carrier diffusion lead to spatially inhomogeneous gain and index profiles. This eventually leads to the formation of light filaments characterised by low gain and high induced refractive index values. Via carrier depletion and self-focusing the filament provides itself with a self-induced waveguide. At the same time the carrier injection leads to the formation of high-gain regions surrounding the filament. The neighbouring regions of higher gain and low index then give rise to new filaments which after some time destabilise the previous filament via the nonlinear transverse coupling. These processes lead to a continuing transverse migration of filaments in space and time. The complexity of the transverse dynamics thereby increases with increasing stripe width.

4.4.2 *Influence of disorder: the spatially inhomogeneous quantum dot ensemble*

In 'real' quantum dot systems spatial inhomogeneities exist: the physical properties of the dots, i.e. size, energy levels and dipole matrix elements, may vary from dot to dot. This has a strong impact on the spatial beam quality, spectral lineshape and spatio-temporal coherence. In our model the consideration of the spatial dependence of the physical quantities allows the self-consistent inclusion of spatial inhomogeneities that exist in real lasers systems. These spatial fluctuations in size and energy levels of the quantum dots and irregularities in the spatial distribution of the quantum dots in the active layer are simulated via statistical methods. Here, we reveal the influence of spatial disorder on the spatio-temporal dynamics. A fundamental analysis of the influence of disorder on gain and spectra is done in Chapter 3. In the following, we show examples of quantum dot lasers with a different degree of spatial disorder. Without loss of generality we will assume a Gaussian distribution in the values of the transition energies, dipole matrix elements and scattering rates. We set the full width at half maximum (FW HM) of the distributions to 2%, 4% and 6%. In the examples we will assume identical fluctuation amplitudes for all parameters. We would like to note, however, that our model allows, in principle, a separate variation of the individual parameters.

As an example, Fig. 4.8 shows at $t = 10, 25$ ns and $10{,}5$ ns (after the startup) snapshots of the distribution of light fields in the active area of a quantum dot laser (operated at approximately $3 \times$ threshold

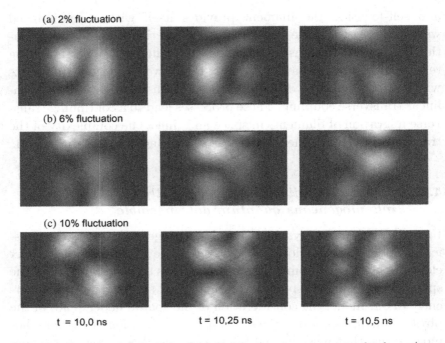

(a) 2% fluctuation

(b) 6% fluctuation

(c) 10% fluctuation

t = 10,0 ns　　　　　t = 10,25 ns　　　　　t = 10,5 ns

Fig. 4.8.　Snapshot of the light field distribution in a quantum dot laser ($w = 50\,\mu$m, $L = 2\,$mm) with spatially varying dot parameters. The amplitude of the (Gaussian shaped) fluctuations is 2%, 6% and 10%.

current density). Its width and cavity length are $50\,\mu$m and $2\,$mm, respectively. In the first example (Fig. 4.8(a)) the dot parameters deviate only slightly from their average values (2% variance) while the second (b) and third (c) plot refer to the situation where the variance of the parameter values of the spatially distributed quantum dots (i.e. their size, dipole matrix elements, energy-levels) is 6% and 10%, respectively. The intensity distributions show characteristic spatial modulations. These modulations change on a picosecond time scale. They originate from dynamic interactions between the light fields and the dot carriers ranging from the femtosecond time scale (in the case of microscopic carrier scattering) up to the picosecond and nanosecond time scales (reflecting the resonator round trip time of the propagating light fields and the slow build-up and decay of the spatial carrier density).

Local variations in the carrier scattering dynamics and optical transitions arising at the individual dots are via dynamic carrier capture and escape, intra-dot level dynamics and the light fields dynamics transformed into characteristic interaction lengths like the coherence length. In combination with light diffraction, propagation of the light fields and the spatially dependent interaction of light with dot carriers this leads to characteristic spatial and temporal modulations and to the formation of dynamic optical patterns. With increasing fluctuation amplitude a characteristic filament structure evolves. The snapshots clearly reveal that the dynamics of these filaments is much more complex than in the case of the 'ideal' quantum dot ensemble (compare Fig. 4.7). In addition, the filaments exhibit a characteristic substructure. The individual level energies lead to locally varying transition energies and frequencies that contribute to the spatial and spectral properties of the propagating light fields. Furthermore, the dynamics of the carrier relaxation processes (phonon emission or absorption and interaction with carriers and phonons of the wetting layers) depends on the energy differences of the levels involved and thus is also directly affected by the spatially varying dot parameters. In combination, these effects cause the dynamic characteristic filament structures that can be seen in Fig. 4.8. Due to the coupling and interplay of spatial with temporal degrees of freedom such material inhomogeneities thus affect both the transverse nearfield distribution as well as the dynamics of the light fields. This will eventually influence relevant quantities such as beam quality and spectra.

Generally, the spectral width of each longitudinal mode is determined by a large variety of physical effects. The characteristic times of induced and spontaneous recombination define a lower limit for a laser linewidth. The real spectral width, however, is significantly broadened. The transverse extension of the structure (laser, amplifier, waveguide, etc.) leads to a characteristic transverse migration of the light fields that is determined by the dynamic interplay of light diffraction, dynamic self-focusing as well as carrier scattering and relaxation. This leads to the formation of a group of transverse modes arising for each longitudinal mode. The number thereby depends on

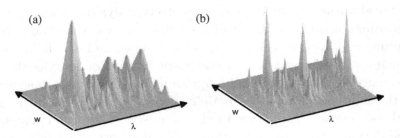

Fig. 4.9.   Spatially resolved emission spectra of quantum dot lasers with Gaussian fluctuations with ampltiude (a) 6% (*left*) and (b) 10% (*right*).

the transverse extension of the active area. In addition, the transverse dynamics is determined by inhomogeneous broadening resulting e.g. from the spatially varying quantum dot parameters (see Chapter 3). They affect both the coherent light–matter coupling (via the carrier dependence of the generation rate) and the incoherent processes such as carrier–carrier and carrier–phonon interactions. Figure 4.9(a)–(b) shows spatially resolved emission spectra corresponding to 4.8 with (a) 6% and (b) 10% fluctuations. The figure clearly demonstrates that an increase in fluctuation amplitude affects both the spectral as well as the spatial degree of freedom. The spatially varying transition energies lead to significant broadening of the emission spectra. This originates from the variance in transition energy and (indirectly) the intra-dot and inter-dot scattering dynamics. The light propagation and diffraction in combination with carrier scattering and relaxation not only lead to a coupling of longitudinal and transverse degrees of freedom but also to a coupling of spectral and transverse dynamics. As a result, the spatially resolved emission spectra show inhomogeneities in both the transverse dimension (i.e. over the lateral extension of the quantum dot broad-area device) and the spectrum.

4.4.3   *Light fluctuations and mode competition in quantum dot cavities*

The large potential of modulated quantum dot laser systems for telecommunications applications has motivated many experimental and theoretical investigations on the characteristics of quantum dot lasers in an external cavity. The external cavity thereby allows us to investigate and to control the fundamental processes of photon emission and interactions. Originally designed to narrow the natural laser linewidth of external cavity lasers, it soon became interesting since they show — for some parameter regimes — characteristic chaotic regimes and instabilities. Today, it is well-known from experiments on bulk and quantum well lasers that feedback realised by the external mirror can induce instabilities and can seriously degrade laser performance. This may lead to a increase in noise and to a broadening of the laser linewidth leading to coherence collapse (Petermann, 1995).

Although they have been studied for some time, semiconductor lasers with delayed optical feedback have recently generated considerable renewed interest (Abdulrhmann *et al.*, 2003; Erzgräber *et al.*, 2006; Fischer *et al.*, 2004; Heil *et al.*, 2001; Mandre *et al.*, 2003; Masoller, 2002; Schikora *et al.*, 2006; Verheyden *et al.*, 2004). The action of the external cavity may be characterised by two parameters: the reflectivity of the external mirror and the external round trip time. Experiments (Carr *et al.*, 2001; Masoller, 2002; Simmendinger *et al.*, 1999) and theoretical studies (Carr *et al.*, 2001; Masoller, 2002; Simmendinger *et al.*, 1999) have demonstrated that the application of delayed optical feedback to bulk or quantum well semiconductor lasers may lead to controlled dynamics or instabilities, depending on the particular value of the reflectivity of the external mirror and depending on its distance from the laser. In the short cavity regime (i.e. external cavities with a few cm length) the formation of regular pulse packaging forming a low-frequency state combined with fast regular intensity pulsations has been found (Heil *et al.*, 2001). This could be explained by the characteristic dynamics of the system around the steady state solution in the phase space. Theoretical studies (Huyet *et al.*, 2000) have discussed the role of low-frequency fluctuations in semiconductor lasers and the influence of coexisting modes has been highlighted in (Buldu *et al.*, 2002). Experiments have shown (Huyet *et al.*, 1998; Vaschenko *et al.*, 1998; Wallace *et al.*, 2000) that the multi-mode dynamics is of particular importance in the regime where low-frequency fluctuations (LFF) occur (i.e. dropouts arising at injection currents close to threshold and moderate feedback levels). It was demonstrated that in the LFF regime an excitation of longitudinal side modes can occur near power dropouts for both frequency-selective (Giudici *et al.*, 1999) and non-selective optical feedback (Huyet *et al.*, 1998; Vaschenko *et al.*, 1998). Further computational studies (Fischer *et al.*, 2000; Yousefi *et al.*, 2001) have shown that external optical frequency filters may allow us control over several dynamical attractors and to stabilise the modes (Green and Krauskopf, 2006).

Not surprisingly, most theoretical and experimental studies done so far concentrate on bulk or quantum well laser. With improvements in quantum dot material technology and the large field of applications of active nano-structures the application of the external cavity concept to quantum dot lasers has recently attracted attention. The nano-structuring thereby leads to a completely different behaviour. The most pronounced difference is the feedback sensitivity (O'Brien *et al.*, 2003; 2004; Su *et al.*, 2003). The two properties that play a key role for the reduction in feedback sensitivity (Helms and Petermann, 1990) are (1) the strongly damped relaxation oscillations (Malić *et al.*, 2006; O'Brien *et al.*, 2004) and (2) an alpha-factor that is — on average — relatively low (O'Brien *et al.*, 2003; Su *et al.*, 2003) but may nevertheless show complex singularities in space and time (Schikora *et al.*, 2006). The damping of the relaxation oscillation thereby is determined by the long length (necessary to guarantee a sufficient gain) and the characteristic dot-wetting layer hole burning (Uskov *et al.*, 2004). This effect will generally limit the maximum modulation bandwidth in modulated laser systems. However, this will be irrelevant in unmodulated devices.

The resistance to feedback instabilities can be advantageous for applications since they enable the operation without the use of optical isolators. Furthermore, their higher stability under the influence of delayed optical feedback makes quantum dot laser interesting candidates to detect and investigate individual dynamic regimes ranging from irregular power drop-outs and periodic pulsations to chaotic behaviour. Recent investigations (Sciamman *et al.*, 2004) have demonstrated that quantum dot lasers may show feedback-induced instabilities that only exist for small alpha factor and vanish if alpha increases. Thereby, the level structure of the dots have an impact on the over-all behaviour. For example, power dropouts in the ground states and intensity bursts in the excited state could be observed in quantum dot semiconductor lasers eimitting in both ground and excited states (Viktorov *et al.*, 2006). The clear route to chaos that can be seen in quantum dot lasers is difficult to observe in quantum well lasers since these devices are much more unstable when exposed to feedback (Carroll *et al.*, 2006). It could be theoretically shown that this behaviour is the result of a relatively low but nonzero line-width enhancement factor and strongly damped relaxation oscillations (Huyet *et al.*, 2004).

The growing relevance of quantum dot lasers in external cavities requires the derivation of suitable theoretical models that realistically take into account the respective material properties and cavity design. Two models for the study of delay-induced temporal instabilities are commonly used: the Lang and Kobayashi model (Lang and Kobayashi, 1980) and the Maxwell–Bloch delay equations (Gehrig *et al.*, 2007). In the Lang and Kobayashi model the laser is described with rate equations for the complex amplitude of the electric field and for the carrier density. Direct influence of spatial effects is disregarded. The optical feedback is considered by adding a term in the field equation corresponding to the re-injection of parts of the emitted field (delayed by an external cavity round-trip). In dimensionless form these equations read:

$$\frac{\partial E}{\partial t} = \kappa(1 + i\alpha)(N - 1)E + \gamma e^{-i\phi_0} E(t - \tau),$$

$$\frac{\partial N}{\partial t} = -\frac{1}{T}(M - J + |E|^2 N). \tag{4.1}$$

In Eq. (4.1) E is the complex amplitude of the light field, N is the carrier density. κ is the field decay rate, T is the carrier decay time and α is the linewidth enhancement factor. The feedback level is represented by γ, J is the pumping parameter; ϕ_0 is the feedback phase when the laser emits at the solitary laser frequency ω_0 and τ is the external cavity round trip time (with $\tau = \phi_0/\omega_0$).

For a simulation of quantum dot structures the above equations can be extended to include the coupling of the dot ensemble with

the embedding quantum well material. The resulting equations read
(Huyet *et al.*, 2004):

$$\frac{\partial E}{\partial t} = -\frac{E}{2\tau_s} + \frac{\Gamma g_0 N_d}{d}(2\rho - 1)E + i\frac{\delta\omega}{2}E + \frac{\gamma}{2}E(t - \tau),$$

$$\frac{\partial \rho}{\partial t} = -\frac{\partial \rho}{\partial \tau_d} - g_0(2\rho - 1)|E|^2 + F(N, \rho),$$

$$\frac{\partial N}{\partial t} = J - \frac{N}{\tau_n} - 2\,N_d R_{cap}(1 - \rho) - R_{esc}\rho, \qquad (4.2)$$

where N is the carrier density in the well and ρ is the occupation
probability in a dot. τ_s is the photon lifetime, τ_n and τ_d are the
carrier lifetime in the well and the dot respectively. N_d is the two-
dimensional density of dots and J denotes the pump term. The factor g_0 is given by the product of cross section of the interaction
of photons with the dot carriers and group velocity, $g_0 = \sigma_0 v_{gr}$.
Γ is the confinement factor and d is the thickness of the dot layer.
The feedback strength and the delay time are given by γ and τ,
respectively. The last two terms in the last line of Eq. (4.2) describe
the rate of exchange of carriers between the well and the dots. The
Lang–Kobayashi model describes the characteristics of many feedback schemes in a realistic way. However, it fails to describe complex carrier effects such as intra-dot scattering between various dot
levels via carrier–phonon scattering leading to a complex multi-level
dynamics affecting the short time dynamics of quantum dot systems.
Furthermore, it does not consider the coexistence and interplay of
longitudinal modes, which is of high importance in many quantum
dot cavity systems.

The Maxwell–Bloch equations on the other hand explicitly
take into account the spatio-temporal nature of the light field and
charge carrier dynamics. In order to include mode-competition we
here use a multi-mode approach (see also Chapter 2.) The multi-
mode Maxwell–Bloch delay equations consider the complex multi-
mode dynamics and the nonlinear dynamic coupling of the light
field dynamics with the microscopic multi-level carrier ensemble. In
their extended form including the relevant properties of a quantum

dot ensemble these equations are composed of the multi-mode Bloch equations presented in Chapter 2 and the following equations describing the dynamics of the light fields propagating in forward and backward directions within the laser:

$$\frac{\partial}{\partial t}\boldsymbol{E}^{\pm} \pm \frac{\partial}{\partial z}\boldsymbol{E}^{\pm} = iD_p\frac{\partial^2}{\partial x^2}\boldsymbol{E}^{\pm} - i\eta\boldsymbol{E}^{\pm} + \Gamma\boldsymbol{P}^{\pm}_{(0)} + \frac{k}{\tau_{in}}\boldsymbol{E}^{\pm}(t - \tau_{ext}).$$

(4.3)

The interaction of the laser-internal light fields with the fields re-entering the active area after a roundtrip in the external resonator is included via the last term in Eq. (4.3); $\tau_{in} = 2n_a L/c$ and $\tau_{ext} = 2L_{ext}/c$ are the internal round trip time and external round trip time, respectively, c is the speed of light, n_a is the refractive index of the active medium and L is the laser cavity length. The dimensionless parameter k takes into account the reflectivity of the external mirror (R_0) and the laser facet mirror reflectivity (R_1). The equations for the carriers and dipoles are identical to the equations in the general quantum dot Maxwell–Bloch theory.

In the following, we concentrate in our discussion of the influence of an external cavity on *spatio-temporal* effects. We will thus show selective results on the influence of feedback on the spatio-temporal light field dynamics of quantum dot lasers. For a systematic parameter variation and discussion of different feedback regimes we refer the reader to the literature (Huyet *et al.*, 2004).

As an example, Fig. 4.10 shows the emission dynamics of a quantum well (a), (b) and a quantum dot laser (c), (d) without (a), (c) and with (b), (d) delayed optical feedback in a temporal range of 10 ns. The feedback has been realised by a delay line of 0.01 ns and a flat mirror with a reflectivity of $R_{xt} = 0.05$. The light field dynamics of the quantum well laser is to a higher degree determined by a transverse migration of the light fields. This is a direct consequence of the large width (here: 50 μm) of the laser cavity. Under the influence of optical feedback (Fig. 4.10(b)) the number of modes contributing to the dynamics is reduced. A thorough optimisation of delay line and feedback can then lead to a systematic control over the modes. It could be shown (Gehrig *et al.*, 2007) that a suitable spatial

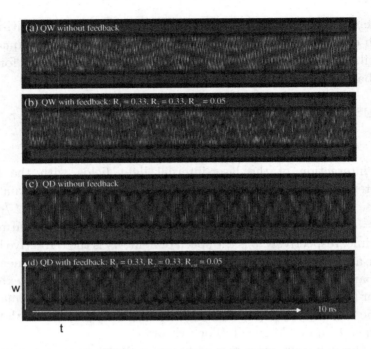

Fig. 4.10. Spatio-temporal dynamics of the light fields emitted at the output
facet of a quantum well (a), (b) and a quantum dot laser (c), (d) without (a), (c)
and with (b), (d) delayed optical feedback in a temporal range of 10 ns.

structuring of the feedback (using a curved mirror) and inclusion of
a spectral filter may allow the over-all control of spatial, spectral and
temporal degrees of freedom leading to single-mode operation. The
quantum dot laser (Fig. 4.10(c),(d)) is less sensitive to delayed optical
feedback. As a consequence, the spatio-temporal dynamics of the system with (c) and without (d) feedback show a very similar behaviour.
We would like to note, however, that — as stated above — a reduced
but non-vanishing influence still exists. Figure 4.11 shows for a delay
line of 0.1 ns the dynamics of the light fields for a reflectivity of the
external mirror of (a) $R_{ext} = 0$, (b) $R_{ext} = 0.05$ and (c) $R_{ext} = 0.5$.
For such a long delay (corresponding to an external resonator with
a propagation length of 3 cm) the simulation reveal comparatively
small changes in the light field dynamics. The situation is changed,
when we consider shorter cavities. Figure 4.12 clearly demonstrates

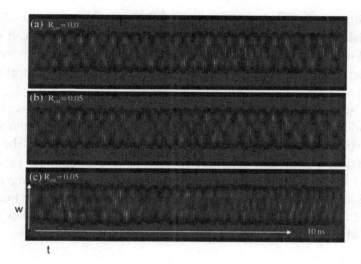

Fig. 4.11. Spatio-temporal dynamics of the light fields emitted at the output facet of quantum dot laser (1 mm length, 50 μm width) with a delay line of 0.1 ns and a feedback strength of $R_{ext} = 0$, (b) $R_{ext} = 0.05$ and (c) $R_{ext} = 0.5$.

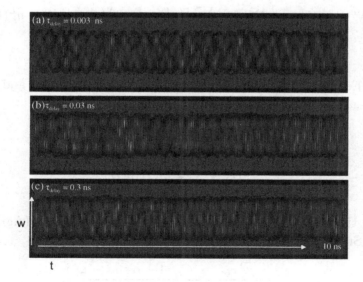

Fig. 4.12. Spatio-temporal dynamics of the light fields emitted at the output facet of quantum dot laser (1 mm length, 50 μm width) with a delay line of (a) 0.003 ns, (b) 0.03 ns and (c) 0.3 ns. The reflectivity of the external mirror has been set to $R_{ext} = 0.05$.

that shorter cavities lead to a stronger influence on the over-all light field field dynamics. This is of high importance for high-speed quantum dot laser systems that use either a segmented contact scheme or short external cavities to generate ultrafast and ultrashort pulse sequences (see also Chapter 7).

As an outlook for future work, we will now present a theory that may additionally be used to fundamentally study the influence of a laser cavity on fluctuation and noise properties. Generally, the noise properties in e.g. bulk or quantum well laser systems can be investigated on the basis of a Langevin approach. The resulting model equations consist of rate equations for photon numbers $n_i(t)$ of each mode i and rate equations for the excited carriers N^c (Becher *et al.*, 1998; Marin *et al.*, 1995):

$$\frac{\partial}{\partial t} n_i(t) = \left[g_i^L(t) - g_i^{NL}(r) - \gamma_i^c \right] n_i(t) + g_i^L(t) + S_i(t) + G_i(t)$$
$$+ g_i(t) + f_i(t),$$

$$\frac{\partial}{\partial t} N^c(t) = \Lambda - \gamma_{sp} N^c(t) - \sum_i \left[\left(g_i^L(t) - g i^{NL}(t) \right) n_i(t) + g_i^L(t) \right]$$
$$+ \Gamma_p(t) + \Gamma_{sp}(t) + \Gamma(t). \tag{4.4}$$

In Eq. (4.4), the terms $g_i^L(t)$ and $g_i^{NL}(t)$ are the linear and nonlinear gain, respectively, given by

$$g_i^L(t) = \frac{\beta_i}{\tau_{sp}} N^c(t)$$

$$g_i^{NL}(t) = \sum_j \epsilon_{ij} g_j^L(t) n_j(t), \tag{4.5}$$

where the nonlinear coupling factor ϵ_{ij} can be written as (Agrawal, 1987; Su *et al.*, 1990)

$$\epsilon_{ij} = \frac{\mu^2 \omega_0}{2\epsilon_0 n n_g \hbar \gamma_c^{ave} \gamma} \frac{C_{ij}}{C_j} \frac{1 + \alpha\tau(\omega_i - \omega_j)}{1 + (\tau(\omega_i - \omega_j))^2}. \tag{4.6}$$

β_i is the spontaneous emission factor into the corresponding mode and τ_{sp} denotes the carrier lifetime due to spontaneous emission. The photon decay rate for each mode can be written as $\gamma_i = \gamma^{res} + \gamma_i^{loss}$ where γ^{res} denotes the decay rate due to ouput coupling losses (which we assume to be equal for all modes) and γ^{loss} describes the decay due to internal losses. Thereby γ_{loss} can be chosen to be different for the individual modes to consider a specific mode suppression ratio in a given experimental setup. $S_i(t)$, is a term describing self-saturation as defined in (Marin *et al.*, 1995): $\gamma_i^{-1} S_i(t) = -s_i(t) \left[\langle n_i \rangle \, \gamma_i / \Lambda \right] \delta n_i(t)$, with a self-saturation parameter s_i, the pumping rate Λ, and the photon number fluctuation $\delta n_i(t) = n_i(t) - \langle n_i \rangle$, where $\langle n_i \rangle$ is the mean value of the photon number. $G_i(t)$, $g_i(t)$ and $f_i(t)$ are the Langevin noise terms due to stimulated emission, internal losses and output coupling, respectively. The terms $\Gamma_p(t)$, $\Gamma_{sp}(t)$ and $\Gamma(t)$ are the Laengevin noise terms for pump noise, spontaneous emission noise, and stimulated emission noise, respectively. The correlation functions for the Langevin terms can be found in (Inoue *et al.*, 1992). The nonlinear coupling factor ϵ_{ij} depends on the transition dipole moment μ, the lasing mode frequencies ω_i, the refractive index n, and the group index n_g. The intraband carrier relaxation rates $\gamma_c^{av} = \gamma_e \gamma_h / (\gamma_e + \gamma_h)$ is an average of the corresponding electron (e) and hole (h) rates, γ is the decay rate of the polarisation. The factor C_{ij}/C_j describes the spatial overlap of the contributing modes. α is the α factor and τ is the polarisation decay time if the coupling and nonlinearity originates from pulsations of the excited carrier population (neglecting intraband effects) (Becher *et al.*, 1998).

The above approach can be further extended to include the coupling between the number of electrons (e), holes (h) in the dots and in the wetting layer (wl) states as well as the loss rate due to an external resonator:

$$\frac{\partial}{\partial t} n_i(t) = \left[g_i^L(t) - g_i^{NL}(r) - \gamma_i^c \right] n_i(t) + g_i^L(t) + S_i(t) + G_i(t)$$
$$+ g_i(t) + f_i(t),$$

$$\frac{\partial}{\partial t} N^{e,h}(t) = \Lambda - \gamma_{sp} N^{e,h}(t) - \sum_i \left[\left(g_i^L(t) - g i^{NL}(t) \right) n_i(t) + g_i^L(t) \right]$$
$$+ \Gamma_p(t) + \Gamma_{sp}(t) + \Gamma(t) - n_{QD} \gamma_{QD-WL} N^{e,h}(t),$$

$$\frac{\partial}{\partial t} N^{WL}(t) = \Lambda^{WL} - \gamma_{sp} N^{WL}(t) + \Gamma_p^{WL}(t) + \Gamma_{sp}^{WL}(t)$$

$$+ \gamma_{QD-WL} N^e(t) + N^h(t). \tag{4.7}$$

The last equation in (4.8) describes the dynamics of the carriers in the wetting layer (WL). The occupation of the wetting layer states may (similarly to the dot system) be changed by the pumping of the wetting layer states (Λ^{WL}), spontaneous emission ($\gamma_{sp} N^{WL}$) and corresponding noise terms ($\Gamma_p^{WL}, \Gamma_{sp}^{WL}$). We thereby neglect induced emission and absorption processes in the wetting layer. γ_{QD-WL} describes the scattering processes between wetting layer and dot states, n_{QD} is a geometrical factor scaling the density of wetting layer carriers to the dot density. The loss rate in an external cavity can be included in the output coupling rate γ_i^c, which may also depend on the mode number i.

These equations now allow a fundamental study of the influence of delayed optical feedback and the influence of co-exisiting modes on the dynamics and noise properties (e.g. photon statistics). For a more detailed inclusion of the influence of the external cavity (including the phase dynamics) we have to consider light fields instead of photon numbers. The corresponding equations then read:

$$\frac{\partial}{\partial t} E_i(t) = \left[g_{E,i}^L(t) - g_{E,i}^{NL}(r) - \gamma_{E,i}^c \right] E_i(t) + g_{E,i}^L(t) + S_{E,i}(t)$$

$$+ G_{E,i}(t) + g_{E,i}(t) + f_{E,i}(t) + r_{ext} E_i(t - \tau_{ext}),$$

$$\frac{\partial}{\partial t} N^{e,h}(t) = \Lambda - \gamma_{sp} N^{e,h}(t) - \sum_i [(g_i^L(t) - g_i^{NL}(t)) n_{ph} |E_i(t)|^2$$

$$+ g_i^L(t)] + \Gamma_p(t) + \Gamma_{sp}(t) + \Gamma(t) - n_{QD} \gamma_{QD-WL} N^{e,h}(t),$$

$$\frac{\partial}{\partial t} N^{WL}(t) = \Lambda^{WL} - \gamma_{sp} N^{WL}(t) + \Gamma_p^{WL}(t) + \Gamma_{sp}^{WL}(t)$$

$$+ \gamma_{QD-WL}(N^e(t) + N^h(t)), \tag{4.8}$$

where r_{ext} denotes the feedback strength and τ_{ext} the delay time in the external cavity. Please note that all terms in Eq. (4.8) referring to fields instead of photon numbers have to be rescaled compared to the corresponding expressions in Eq. (4.7), n_{ph} is the factor required to calculate photon numbers from intensities $|E_i(t)|^2$.

4.5 Conclusion

The results of this chapter demonstrate that nano-structured photonic semiconductor devices exhibit a highly nonlinear and complex light field and carrier dynamics that depends on a hierarchy of physical processes occurring on different time scales. Simulations on the basis of the quantum dot Maxwell–Bloch equations explicitly taking into account the transverse and longitudinal space coordinates allow a realistic representation of typical spatio-spectral properties. The simulation results visualise the dynamics of the light fields and carriers and provide insight into the coupling and interplay of spatial and temporal degrees of freedom. Typical properties of quantum dot devices such as spatial inhomoeneity and dynamic coupling of dot states and wetting layer states could be found to play key roles in the spatial and spectral emission characteristics of the lasers. This highlights the importance of growth of QD material with low fluctuations in size and energy levels (in the regime of few percentage). The performance of QD lasers has been shown to be superior to that of quantum well lasers with respect to feedback sensitivity. Our results show that a careful design of the QD laser geometry and material may be crucial for applications. Using the details of a given material system and laser geometry our space-time model thus opens the way to explore and predict the physical properties of state-of-the-art lasers and to contribute to the design and the development of optimised active nano-structures.

References

Abdulrhmann, S.G., Ahmed, M., Okamoto, T., Ishimori, W. and Yamada, M., *IEEE J. Sel. Top. Quant.* **9**, 1265–1274, (2003).

Adachihara, H., Hess, O., Abraham, E., Ru, P. and Moloney, J.V., *J. Opt. Soc. Amer. B* **10**, 658–665, (1993).

Agrawal, G.P., IEEE Journal of Quantum. Electron. *IEEE J. Quantum Elect.* **23**, 860–868, (1987).

Becher, C., Gehrig, E. and Boller, K.-J., *Phys. Rev. A* **57**, 3952–3960, (1998).

Buldu, J.M., Rogister, F., Trull, J., Serrat, C., Torrent, M.C., Garcia-Ojalvo, J. and Mirasso, C.R., *J. Opt. B* **4**, 415–420, (2002).

Burkhard, T., Ziegler, M.O., Fischer, I. and Elser, W., *Chaos Solitons Fract.* **10**, 845–850, (1999).

Carr, T., Pieroux, D. and Mandel, P., *Phys. Rev. A* **63**, 33817–33831, (2001).

Carroll, O., O'Driscoll, I., Hegarty, S.P., Huyet, G., Houlihan, J., Viktorov, E.A. and Mandel, P., *Opt. Express* **14**, 10831–10837, (2006).

Erzgräber, H., Krauskopf, B., Lenstra, D., Fischer, A.P.A. and Vemuri, G., *Phys. Rev. E.* **73**, 55201–55204, (2006).

Fischer, A.P.A., Andersen, O.K., Yousefi, M., Stolte, S. and Lenstra, D., *IEEE J. Quantum Elect.* **36**, 375–384, (2000).

Fischer, A.P.A., Yousefi, M., Lenstra, D., Carter, M.W. and Vemuri, G., *Phys. Rev. Lett.* **92**, 023901, (2004).

Gehrig, E. and Hess, O., *Dynamics of high-power diode lasers* in *Topics in Applied Optics*, Diehl, R., (Ed.), Heidelberg, Springer Verlag, (2000), vol. 3, pp. 55–82.

Gehrig, E. and Hess, O., *Spatio-Temporal Dynamics and Quantum Fluctuations in Semiconductor Lasers.* Springer-Verlag, Berlin, (2003).

Gehrig, E. and Hess, O., *Appl. Phys. Lett.* **86**, 203116, (2005).

Gehrig, E., Gaciu, N. and Hess, O., Control of Broad-Area Laser Dynamics via Delayed Optical Feedback, invited chapter to 'Handbook of Chaos Control', Schoell, E. and Schuster, H.G., (Eds.), Wiley (2007).

Giudici, M., Giuggioli, L., Green, C. and Tredicce, J.R., *Chaos Solitons Fractals* **10**, 811–818, (1999).

Green, K. and Krauskopf, B., *Optics Communications* **258**, 243–255, (2006).

Heil, T., Fischer, I., Elsäßer, W. and Gavrielides, A., *Phys. Rev. Lett.* **87**, 243901, (2001).

Helms, J. and Petermann, K., *IEEE J. Quantum Elect.* **26**, 833–836, (1990).

Huyet, G., Balle, S., Giudici, M., Green, C., Giacomelli, G. and Tredicce, J.R., *Opt. Commun.* **149**, 341–347, (1998).

Huyet, G., Porta, P.A., Hegarty, S.P., McInerney, J.G. and Holland, F., *Opt. Commun.* **180**, 339–344, (2000).

Huyet, G., O'Brien, D., Hegarty, S.P., McInerney, J.G., Uskov, A.V., Bimberg, D., Ribbat, C., Ustinov, V.M., Zhukov, A.E., Mikhrin, S.S., Kovsh, A.R., White, J.K., Hinzer, K. and Spring Thorpe, A.J., *Phys. Stat. Sol.* (*a*) **201**, 345–352, (2004).

Inoue, S., Ohzu, H., Machida, S. and Yamamoto, Y., *Phys. Rev. A* **46** 2757–2765, (1992).

Lang, R. and Kobayashi, K., *IEEE J. Quantum Elect.* **16**, 347–355, (1980).

Malič, E., Ahn, K.J., Bormann, M.J.P., Hövel, P., Schöll, E., Knorr, A., Kuntz, M. and Bimberg, D., *Appl. Phys. Lett.* **89**, 101107, (2006).

Mandre, S.K., Fischer, I. and Elsäßer, W., *Opt. Lett.* **28**, 1135–1137, (2003).

Mandre, S.K., Fischer, I. and Elsäßer, W., *Opt. Comm.* **244**, 355–365, (2005).

Marciante, J.R. and Agrawal, G.P., *IEEE J. Quantum Elect.* **33**, 1174–1179, (1997).

Marin, F., Bramati, A., Giacobino, E., Zhang, T.C., Poizat, J.-Ph., Roch, J.-F. and Grangier, P., *Phys. Rev. Lett.* **75**, 4606, (1995).

Masoller, C., *Physica D* **168–169**, 171–179, (2002).

O'Brien, D., Hegarty, S.P., Huyet, G., McInerney, J.G., Kettler, T., Laemmlin, M., Bimberg, D., Ustinov, V.M., Zhukov, A.E., Mikhrin, S.S. and Kovsh, A.R., *El. Lett.* **39**, 1819, (2003).

O'Brien, D., Hegarty, S.P., Huyet, G. and Uksov, A.V., *Opt. Lett.* **29**, 1072, (2004).

Petermann, K., *IEEE J. Sel. Top. Quant.* **1**, 480–489, (1995).

Schikora, S., Hövel, P., Wünsche, H.-J., Schöll, E. and Henneberger, F., *Phys. Rev. Lett.* **97**, 213902, (2006).

Sciamman, M., Megret, P. and Blondel, M., *Phys. Rev. E* **69**, 046209, (2004).

Simmendinger, C., Preisser, D. and Hess, O., *Opt. Express* **5**, 48–54, (1999).

Su, C.B., Schlafer, J. and Lauer, R.B., *Appl. Phys. Lett.* **57**, 849–851, (1990).

Su, H., Gray, A.L., Wang, R., Newell, T.C., Malloy, K.J. and Lester, L.F., *IEEE Photon. Technolog. Lett.* **15**, 1504, (2003).

Uskov, A.V., Boucher, Y., Le Bihan, J. and McInerney, J., *Appl. Phys. Lett.* **73**, 1499, (1998).

Vaschenko, G., Giudici, M., Rocca, J., Menoni, C., Tredicce, J. and Balle, S., *Phys. Rev. Lett.* **81**, 5536, (1998).

Verheyden, K., Green, K. and Roose, D., *Phys. Rev. E* **69**, 036702, (2004).

Viktorov, E.A., Mandel, P., O'Driscoll, I., Carroll, O., Huyet, G., Houlihan, Y. and Tanguy, Y., *Opt. Lett.* **31**, 2302–2305, (2006).

Wallace, I., Yu, D., Harrison, R.G. and Gavrielides, A., *J. Opt. B* **2**, 447, (2000).

Yousefi, M., Lenstra, D., Vemuri, G. and Fischer, A., *IEE Proc.-Optoelectron.* **148**, 233, (2001).

Chapter 5

Performance and Characterisation: Properties on Large Time and Length Scales

5.1 Introduction

Many application fields require good spatial and spectral quality of the radiation emitted by a quantum dot laser. In the previous chapters we have seen that the light fields emitted in the first nanoseconds show particularly complex dynamics. This is a direct consequence of the large amplitudes in the changes within the charge carrier plasma and in the light fields characterising the typical relaxation oscillations. On longer time scales, i.e. when considering temporal regimes up to several 100 ns, the influence of the typical start-up behaviour decreases and the dynamics is mostly determined by the different length and time scales of the physical processes (e.g. light diffraction, carrier scattering and diffusion, self-focusing) affecting the nonlinear light–matter interaction. In order to model longer time scales in a reasonable computation time we here use the effective two-level multi-mode Maxwell–Bloch approach presented in Chapter 2. This approach takes into account all processes relevant for the laser performance such as, in particular, multi-mode dynamics, spatial dependence of the light field and carrier dynamics, dynamic gain and index changes as well as the amplitude phase coupling (α-factor).

5.2   Spatial and Spectral Beam Quality

For an analysis of the beam quality we temporally average the light fields at the output facet and calculate an average near field. Similarly to our discussion in the previous chapters it is very instructive to compare the emission of a quantum dot laser with the emission of quantum well laser in order to elucidate the advantageous influence of the nano-structured material. Figure 5.1 shows an example of time averaged nearfields of a quantum well (a) and a quantum dot (b) laser. In the simulation the stripe width had been set to 6 μm. Displayed are the distributions for three output power levels. The comparison of the nearfields demonstrates that for this stripe width the quantum dot laser has a Guassian-shaped nearfield distribution, whereas the quantum well laser shows characteristic modulations. The figures prove the strong influence of the transverse degree of freedom in the quantum well laser. These results obtained on the basis of the Maxwell–Bloch equations were in very good agreement with nearfield distributions obtained in an experiment (Gehrig *et al.*, 2004). The suppressed transverse light field dynamics observed in experiments and predicted by the simulation clearly demonstrate

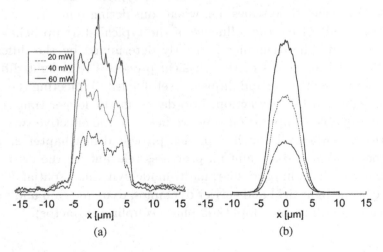

Fig. 5.1.   Nearfield of a quantum well (a) and a quantum dot laser (b) in dependence on output power.

the promising device performance of nano-structures compared to large area lasers and laser-amplifiers which show a strong tendency for filamentation formation (Gehrig *et al.*, 1999; Lang *et al.*, 1994; Marciante and Agrawal, 1996).

In order to quantify the beam quality we have systematically varied the laser width and the injection current of the quantum dot and the quantum well laser. The theoretical beam quality factor M^2 as obtained by the spatio-temporal simulation is depicted in Fig. 5.2 in dependence on stripe width (a) and output power (b) (for the same output power of 20 mW). With increasing stripe width (Fig. 5.2(a)) the transverse degree of freedom gets more important: the characteristic dynamic optical patterns resulting from the dynamic interplay of carrier diffusion and diffraction that could be seen in the dynamics of the light fields (see e.g. Figs. 4.1 and 4.2 in Chapter 4) lead to characteristic dynamic optical patterns that typically lie in the μm-regime. In combination with the dynamic phase changes, this results in a deterioration of the beam quality, i.e. the M^2-parameter of the quantum well and the quantum dot laser increases with increasing stripe width. However, due to the strong localisation of the carriers and the reduced α-factor, the M^2-values of the quantum dot laser are always smaller than the respective values of the quantum well structure. In particular, the quantum dot laser shows a characteristic

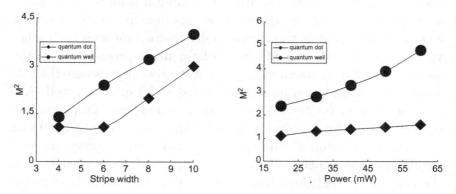

Fig. 5.2.   Beam quality factor (M^2) of quantum well and quantum dot lasers. Shown are the dependence of stripe width (a) and power (b).

threshold near $8\,\mu$m. In this intermediate stripe regime the quantum dot laser is still single-mode whereas $M^2 > 2$ for the quantum well laser. The dependence of M^2 on output power is shown in Fig. 5.2(b) (for a stripe width of $6\,\mu$m). In the quantum well laser an increase in the injection current density not only increases the output power but simultaneously leads to an increase in the M^2-parameter. This is a direct consequence of dynamic carrier diffusion and light diffraction affecting the light fields during their propagation in the laser. In contrast, the quantum dot laser shows almost no dependence on output power. The dependence of M^2 on stripe width and output power can be confirmed by experimental measurements performed on the same devices (Gehrig *et al.*, 2004).

Our numerical results clearly demonstrate that quantum dot lasers have a much better beam quality compared to quantum well lasers of the same geometry. The strong localisation of the carriers in the dots in combination with the reduced amplitude phase coupling thus guarantees a good spatial quality.

In general, the width and the length of the laser cavity determine a number of possible transverse and logitudinal modes. The actual spectra of a given active laser structure, however, depends on both the geometry and material properties. The longitudinal as well as the transverse modes influence each other via the strongly nonlinear coupling of the light fields with the spatio-temporal carrier dynamics in the active medium. Consequently, the temporal behaviour and emission spectra that can be observed in experiments are the result of a complex interplay of transverse modes interacting with each other via the active medium and via the light fields propagating in the laser. Figure 5.3 shows an example of a (temporally averaged) mode spectra of a broad area quantum dot (*top*) and quantum well (*bottom*) semiconductor laser for two injection currents. Displayed are optical spectra for a stripe width of $30\,\mu$m and a cavity length of $L = 3\,$mm. The number and spectral distance $\Delta\omega$ between the longitudinal modes depends via $\Delta\omega = c/(2n_l L)$ on both the length L of the laser cavity and the refractive index n_l of the active laser material. The results show an increase in mode number with increasing carrier injection level. A structuring on nano-scales thereby reduces

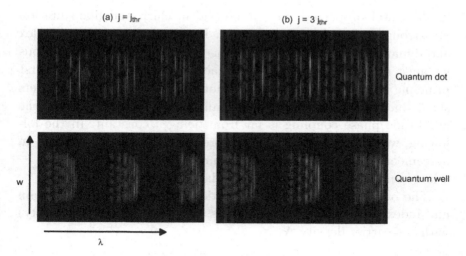

Fig. 5.3. Emission spectra: influence of the current density on the spectral properties of quantum dot (*top*) and quantum well (*bottom*) lasers. The stripe width of both lasers is $30\,\mu$m.

the influence of the transverse degree of freedom. As discussed in Chapter 4 this leads to a reduced transverse dynamics of the light fields. In the emission spectra, this leads to a reduction of the number of transverse modes.

5.3 Dynamic Amplitude Phase Coupling

The amplitude phase coupling as described by the so-called α-factor is a convenient measure to describe the magnitude of the quality and emission characteristics of semiconductor lasers. Since the early days of laser science the α-factor has been used as an important parameter for the classification of laser structures. In the 80s, first measurements revealed that the semiconductor laser linewidth was much broader than the Shawlow Townes limit. Soon after, Henry (1982) theoretically explained the excess linewidth with the coupling of index and gain via the carrier dynamics. Since then, the role of the α-factor has been in the focus of many theoretical and experimental publications (Osinski and Buus, 1987). In semiconductor lasers any fundamental properties such as linewidth, chirp and filamentation are strongly related to the α-factor. But having seen the complex and dynamic light–matter interactions in our space-time simulations we must immediately ask: is the α-factor really a parameter? First principle simulations of InGaAs quantum dot lasers and amplifiers show that in spatially extended quantum dot laser structures the amplitude phase coupling is far from being a constant. In the following, we will see that our theory predicts and demonstrates — in agreement with corresponding experimentally determined values — a large and excitation-dependent variation and scatter.

The α-factor is defined as the ratio of the variation of the gain and index (or real and complex parts of the complex susceptibility) with the carrier density N

$$\alpha = -\frac{4\pi}{\lambda}\frac{dn/dN}{dg/dN} = \frac{d\mathrm{Re}[\chi]/dN}{d\mathrm{Im}[\chi]/dN}, \qquad (5.1)$$

where n is the (induced) refractive index, g the gain and N the carrier density.

Generally, the coupling of gain and index or amplitude and phase strongly depends on the material and cavity design of the active semiconductor medium. Due to their discrete level energies quantum dot laser structures are characterised by a rather symmetric gain and induced index dispersion. One would thus expect, that they have, in principle, a small α-factor. Indeed, values near one have been found for InGaAs/GaAs QD lasers in experiments. Furthermore, it has been demonstrated that the α-factor in QD lasers is much lower and shows less dispersion than in quantum well lasers of identical material and geometry (Ukhanov *et al.*, 2004). However, the measured values of α strongly vary, even in the same type of quantum dot laser from device to device (Schneider *et al.*, 2004). It is thus of high relevance to explore the *dynamic* physical processes that affect the optical properties of quantum dot lasers and eventually lead to corresponding changes in the spatio-temporally averaged value of the α-factor.

The simulations are based on the spatially resolved quantum dot Maxwell–Bloch equations presented in Chapter 2. These equations explicitly consider the full dynamics of carriers and inter-level dipoles. As a consequence, it is principally not required to include an 'artificial' α parameter in the calculations. However, we can extract the value of this parameter via the dynamically calculated real and imaginary part of the polarisation that are via the light fields directly correlated to the susceptibility. In the following we will address the following questions: (i) How does α evolve after switching on the laser? (ii) How does the characteristic spatial disorder of quantum dot ensembles and the light field dynamics affect α? and (iii) How does the ultrafast dynamics feed back onto α?

Figure 5.4 shows (for a quantum dot laser of 1 mm length and 50 μm stripe width) the dynamics of the (spatially averaged) amplitude phase coupling after the start-up of the laser (i.e. switching the current from zero to its final value). Immediately after switching on the current the high non-equilibrium dynamics of the charge carrier plasma and the light fields leads to a high amplitude phase coupling. The spatial average of α shows a relaxation behaviour that originates from the out-of-equilibrium dynamics of light fields and carriers within the first few nanoseconds. During the first nanoseconds the relaxation oscillations slowly approach their quasi-equilbrium. This is directly reflected in the decay in α. It is important to realise that the spatio-temporal average of α that can be measured on longer

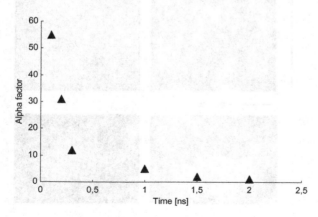

Fig. 5.4.   Dynamics of α in a quantum dot laser ($L = 1$ mm, $w = 50\,\mu$m): spatial average in dependence of time immediately after the switching on of the current.

time scales is not identical to the value that can be observed in the first nanoseconds. It is, in particular, the strong dependence of α on carrier and light field dynamics that is responsible for a characteristic dynamics in α persisting for a few nanoseconds.

In a next step, we investigate the *spatial* variation that is indirectly contained in a spatio-temporal average. As an example, we plot the spatial distribution of the intensity (*left*) and the α-factor (*right*) for the same quantum dot laser ($L = 1$ mm, $w = 50\,\mu$m) for a time $t = 2.5$ ns after switching on the current (Fig. 5.5). The snapshot shows an image (in 2D (a) and 3D (b) view) of dynamic μm-sized light patterns frozen in time. The formation of this pattern originates from the dynamic light–matter coupling characterised by light diffraction, self-focusing and complex carrier scattering. At the same time the spatially varying carrier and intra-dot dipole dynamics leads to the highly complex dynamics and the strong spatial dependence of the α-factor. It is particularly noticeable that moderate changes in the light field and carriers lead to very pronounced patterns and

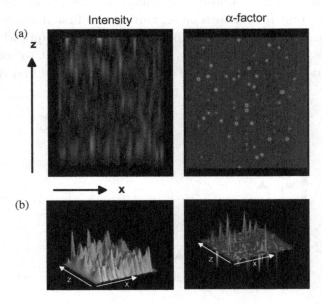

Fig. 5.5.   Snapshots of the intensity (*left*) and α-factor (*right*) in a quantum dot laser ($L = 1$ mm, $w = 50\,\mu$m) at 2.5 ns in (a) 2D and (b) 3D depiction.

spikes in the amplitude phase coupling. α thus is — via its definition — much more sensitive to spatial and temporal variations!
Indeed, typical values for α may, in a given laser structure, spatially
and temporally vary from values between -3 to 3 or even show singularities while the average with respect to space and time may be
considerably lower. Although α may be comparatively small in each
individual dot at a particular moment in time the complex amplitude
and phase dynamics resulting from the characteristic spatially varying material properties (e.g. energy levels and dipole matrix elements
in combination with the dynamic coupling to the counterpropagating
light fields) lead to strong spatial and temporal changes in α. The
α-factor consequently is far from being constant. A systematic variation of the laser geometry and the injection current confirms this
fact. Figure 5.6(a) shows the dependence of α (temporally and spatially averaged) on the width of the laser. Figures 5.6(b) and 5.6(c)
demonstrate the dependence of α on carrier injection for a narrow-
stripe laser (stripe width of $6\,\mu$m) and a broad-stripe laser (transverse
stripe width $50\,\mu$m), respectively.

As discussed in Chapter 4 increasing the width of the active area
leads to a growing influence of the transverse dimension: physical processes such as carrier diffusion, light diffraction and dynamic phase
changes not only result in a notable deterioration of the beam quality (Gehrig *et al.*, 2004) but also in an increase in α (Gehrig and
Hess, 2004). The narrow-stripe quantum dot laser (width of $6\,\mu$m,
Fig. 5.6(b)) is characterised by a slight increase in α with increasing
carrier density. This behaviour originates from increased influence of
the intra dot carrier dynamics characterised by carrier capture and
carrier–phonon interactions coupling the various energy levels within
the dot. As a consequence the carrier induced gain and index dispersion show complex dynamics leading to an increase in α. Increasing
the stripe width (Fig. 5.6(c)) to $50\,\mu$m leads to an additional increase
in complexity: a spatially inhomogeneous dot ensemble (in the example: $18\,$meV) leads to a rise in the amplitude phase coupling as a
result of the increased carrier, gain and index dynamics whereas for
an ideal laser with homogeneous material properties (i.e. identical

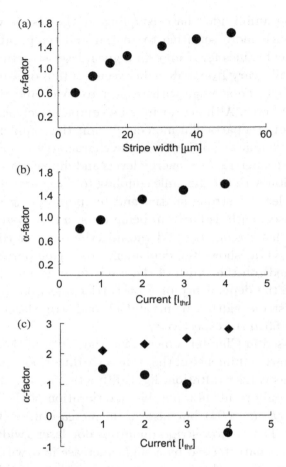

Fig. 5.6.   Dependence of alpha on stripe width (a) and on injection current (b), (c) for a stripe width of 6 μm (b) and 50 μm. Dots and diamonds in (c) represent homogeneous and inhomogeneous dots, respectively.

dot size, matrix elements and uniform dot distribution) the amplitude phase coupling may even decrease with increasing stripe width. This effect can be attributed to strong Coulomb interactions that may shift the characteristic gain and index distributions as well as the resulting α parameter (Smowton, *et al.*, 2004).

In a final step we now investigate the dependence of α on optical excitation. Figure 5.7 shows a snapshot of the intensity (a) and

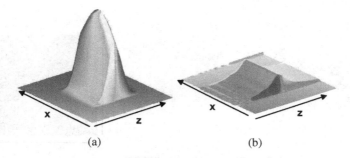

Fig. 5.7. Snapshots of intensity (a) and α-factor (b) during the propagation of a light pulse.

the α factor (b) in the active area of a QD laser amplifier (width 6 μm, length 1 mm) into which a resonant picosecond light pulse has been injected. The injected light pulse leads to a dynamic excitation and relaxation within the charge carrier system. This excitation and relaxation dynamics is directly reflected in the rise and decrease of α (Fig. 5.7(b)). Under these particular conditions, α represents the material response to the optical excitation and more or less mirrors the pulse shape. We would like to note that the specific shape and (spatial as well as temporal) offset of α strongly depends on carrier inversion and the degree of saturation. In strongly saturated amplifiers the spatial distribution of α may be significantly distorted while the propagating pulse is significantly reshaped. Furthermore, in laser amplifiers with large stripe widths, α may show a strong transverse dependence. A particular interesting situation is the injection of an ultrashort pulse (pulse duration <1 ps). In this case the carrier relaxation processes occur on similar time scales leading to dynamic changes in amplitude and phase within the pulse envelope. We thus expect a higher degree of structuring and complexity in α. Indeed, Fig. 5.8 showing the dynamics of α during the passage of a 150 femtosecond light pulse in a quantum dot laser amplifier (stripe width 50 μm) clearly demonstrates a complex dynamics. The figures on the left side visualise the spatially averaged dynamics of α at the output facet for (a) a spatially inhomogeneous and (b) a homogeneous ensemble of QD. The figures on the right show (for the first 30 ps) the

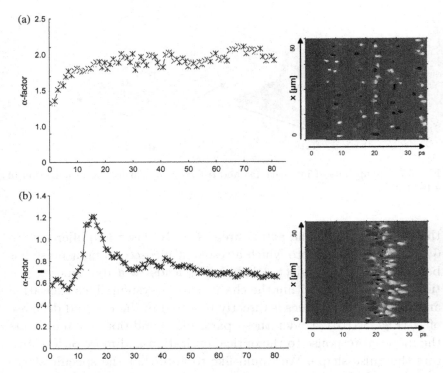

Fig. 5.8.   Dynamics of alpha during the propagation of a light pulse. *Left*: spatial average, *right*: laterally resolved distributions.

respective transversely resolved dynamics of α at the output facet. In the spatially inhomogeneous dot system (Fig. 5.8(a)) the light pulse induces a highly non-equilibrium situation in the charge carrier system. It is the dynamic interplay of counter-propagating light fields and spatially dependent dipole dynamics that leads to the dynamic spatial phase pattern in the active area. As a consequence the non-equilibrium dynamics induced by the light pulse may lead to a long-lived excitation (Fig. 5.8(a)). Although α reaches a quasi-stationary state after approximately 50 ps the laterally resolved calculation of the first 30 ps clearly reveal strong spatial variations and spiking. In the homogeneous dot ensemble (Fig. 5.8(b)) the over-all relaxation of induced gain and index is faster leading to a characteristic rise and decreasing value of α.

In conclusion, our space-time simulations of α clearly prove a strong spatio-temporal dynamics. They provide a key to the interpretation of experimental measurements and give a measure of applicability (and limits) of the α-factor for the classification of active nano-structured laser devices.

5.4 Conclusion

Space-time simulations of typical laser devices allows the prediction and interpretation of relevant properties such as, in particular, spatial and spectral beam quality, feedback sensitivity, homogenous and inhomogenous gain broadening as well as the amplitude-phase coupling. The simulations confirm and predict spatial and spectral emission properties that are relevant in experiments. The performance of QD lasers has been shown to be superior to that of quantum well lasers in the critical feature of the α factor. The Maxwell–Bloch approach provides a fundamental description of the underlying physical processes and guarantees a realistic modelling of the laser-internal amplitude phase coupling. Our results not only predict the average α factor but also reveal the correlation between spatio-temporal light field and carrier dynamics and the complex dynamics of the amplitude phase coupling. They demonstrate that — in spite of the small spatio-temporal average — the α factor in QD lasers shows a complex spatial and temporal dynamics and is far from being constant. Although a measured α-factor (or the average of a calculation) may be appropriate for a first evaluation of the optical properties of a QD laser device one thus has to be aware that the laser-internal amplitude phase coupling is a strongly varying physical quantity affected by complex light–matter interactions and carrier dynamics.

References

Gehrig, E., Hess, O. and Wallenstein, R., *IEEE J. Quantum Elect.* **35**, 320–331, (1999).

Gehrig, E., Hess, O., Ribbat, C., Sellin, R.L. and Bimberg, D., *Appl. Phys. Lett.* **84**, 1650–1652, (2004).

Gehrig, E. and Hess, O., *Appl. Phys. Lett.* **86**, 203116, (2005).

Henry, C.H., *IEEE J. Quantum Elect.* **18**, 259–264, (1982).

Lang, R., Hardy, A., Parke, R., Mehuys, D. and O'Brien, S., *IEEE J. Quantum Elect.* **30**, 658–693, (1994).

Marciante, J.R. and Agrawal, G.P., *IEEE J. Quantum Elect.* **32**, 590–596, (1996).

Osinski, M. and Buus, J., *IEEE J. Quantum Elect.* **23**, 9–29, (1987).

Schneider, S., Borri, P., Langbein, W., Woggon, U., Sellin, R.L., Ouyang, D. and Bimberg, D., *IEEE J. Quantum Elect.* **40**, 1423–1429, (2004).

Smowton, P.M., Pearce, E.J., Scheider, H.C., Chow, W.W. and Hopkinson, M., *Appl. Phys. Lett.* **81**, 3251–3253, (2002).

Ukhanov, A.A., Stintz, A., Eliseev, P.G. and Malloy, K.J., *Appl. Phys. Lett.* **84**, 1058–1060, (2004).

Chapter 6

Nonlinear Pulse Propagation in Semiconductor Quantum Dot Lasers

In a semiconductor quantum dot laser the dynamic interplay of light and matter involves a hierarchy of different time (fs...ns) and length scales (nm ... μm). Indeed, the quantum dot ensemble represents a highly complex medium in which dynamic nonlinear phenomena may occur. Particularly the ultrashort time dynamics in the charge carrier plasma shows a strong dependence on amplitude and spatio-spectral properties of the optical fields. These characteristic effects thus can be best monitored by simulating the propagation and amplification of an ultrashort light pulse. The space-time simulations then directly reveal the underlying physical interactions and allow for a visualisation of the spatio-temporal dynamics. While revealing a whole world of fundamental processes, the simulation of pulse propagation also is of high technological relevance: due to the short intra-dot relaxation processes quantum dot lasers are particularly promising candidates for modulated systems and high-speed applications.

We note that a significant advantage of the approaches presented here is the full inclusion of spatial variations and spectral dependencies represented in the level structure of the dot ensemble. Furthermore, the explicit spatial and temporal integration of the equations does not require a transformation of the variables to the travelling frame of the propagating pulse that is typically done in phenomenological models. Thus, the dynamic interaction of the light field with the nonlinear matter is fully taken into account.

In this chapter we will discuss the spatio-temporal dynamics of an InGaAs quantum dot waveguide with injection of an ultrashort light pulses. Due to the specific physical properties of the active medium such as, for example, discrete energy levels, high gain, low threshold current, low alpha-factor (Ghosh *et al.*, 2002) and high-speed modulation (Hatori *et al.*, 2004) quantum dot based semiconductor optical amplifiers (SOAs) represent particular promising devices for novel optoelectronic components in data storage and telecommunication.

We will base our simulations on the quantum dot Maxwell–Bloch–Langevin equations presented and discussed in Chapter 2. We will concentrate on the dynamic shaping of a pulse and on the propagation time of a pulse signal passing an active quantum dot medium and explore the possibility to control its propagation speed ('slow and fast light'). These properties are a direct result of the ultrashort femtosecond dynamics discussed in Chapter 3. We will see that the microscopic spatio-temporal interaction of the counterpropagating light fields with the microscopic distribution of the electron-hole system and the nonlinear interband polarisation directly leads to a dynamic spatial and spectral shaping of a light pulse propagating in a semiconductor medium.

6.1 Dynamic Shaping of Short Optical Pulses

The dynamic interactions between light and matter have a strong influence on the physical properties of a light pulse propagating in the active area of a quantum dot laser. The physical processes affecting the properties of a light pulse are e.g. spatio-spectral hole burning, carrier heating and thermalisation leading to dynamic spatial and spectral gain saturation, induced index dispersion and self-phase modulation. This is of particular importance for spatially extended inhomogeneous media, where severe nonlinearites and instabilities can emerge (Hess and Kuhn, 1996).

The Maxwell–Bloch–Langevin equations provide a realistic description of the propagation of light pulses in active nanostructured media. The reduction of the spatio-spectral inversion by induced recombination and the dynamics of the interband polarisation representing spatio-spectral gain and index (see Chapter 3) lead to dynamic saturation, self-focusing and diffraction. Together they affect both the amplitude and phase of a propagating light field. In the following we will present results of our numerical modelling of the pulse propagation in quantum dot lasers. Thereby, we will focus on picosecond pulses and on pulses with a duration less than 1 ps.

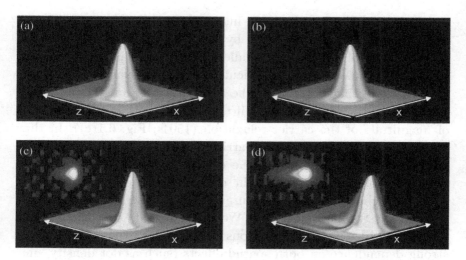

Fig. 6.1.   Snapshot of the intensity in the active area of an optically injected quantum dot laser (width of active laser stripe: 10 μm, cavity length 500 μm) with a duration of (a), (c) 150 fs and (b), (d) 1.5 ps of the injected pulse. Figures (a) and (b) refer to the small signal regime, Fig. (c,d) to the saturation regime.

Figure 6.1 shows snapshots of light pulses propagating in a quantum dot laser amplifier after a propagation length of 500 μm). The figures refer to the small signal regime (Fig. 6.1(a), (b)) and to the saturation regime (Fig. 6.1(c), (d)) for a pulse with a duration (full width at half maximum) of 150 fs (Fig. 6.1(a), (c)) and 1.5 ps (Fig. 6.1(b), (d)), respectively. For a better visualisation we plot only a small area around the pulse and not the entire active area. The characteristic spatial and temporal distortions the light pulse experiences during propagation in an inverted dot medium strongly depend on the duration of the light pulse: in the small signal regime (e.g. small intensity values of the injected light pulses), a light pulse more or less retains its shape during its propagation in an inverted quantum dot laser amplifier, independent of the duration of the injected pulse. However, if we increase the input power level (approaching the saturation regime of the quantum dot waveguide) the dynamic pulse shaping shows a strong dependence on the duration of the pulse. The leading part of the pulse with duration of 1.5 ps (Fig. 6.1(d)) is significantly amplified by the nonlinear interaction with the quantum dot ensemble. The trailing part interacts with the carrier inversion that has already been modified by the leading part. Consequently, the experienced gain is much smaller leading to an asymmetric pulse shape. In this case, it is, in particular, the spatial variation of inversion and gain that determines the amplification and shaping of the pulse. The situation is changed if the pulse duration is in the order of magnitude of the carrier relaxation (150 fs, Fig. 6.1(c)). In that case, the microscopic intradot carrier dynamics and the interaction with carriers and phonons of the embedding medium gain importance. In combination with light diffraction and propagation this leads to a characteristic curvature of the pulse front and modulations in the trailing pulse part. We thus conclude that the amount of spatial and temporal distortions a light pulse experiences show a strong dependence on both spatial effects (such as dot density, uniformity of the dot distribution, spatial fluctuations) and microscopic 'spectral' effects determined by the characteristic relaxation times and the physical properties of the individual dots.

6.2 Nonlinear Femtosecond Dynamics of Ultrashort Light Pulses

The theoretical analysis of the femtosecond dynamics during the propagation of ultrashort light pulses is a typical model example suitable for the study of the influence of microscopic carrier dynamics and nonlinear effects on the properties of the propagating signal. We will show that physical mechanisms such as spatio-spectral level hole burning and ultrafast gain and index dynamics not only affect the spectrum and the (temporal) shape of a light pulse but additionally its speed.

6.2.1 *Self-induced propagation control: tunable propagation speed*

As discussed in Chapter 3 the carrier dynamics in quantum dot lasers is directly correlated to a complex gain dynamics and index dispersion that eventually determine amplitude and phase — and consequently also the speed — of a propagating light pulse. For an investigation of this dynamic speed control we will in the following simulate situations where an ultrashort pulse propagates in a quantum dot laser and calculate the propagation time. We will reveal the dependence of the propagation time on injection current density and pulse energy. Furthermore, our simulations will allow for a prediction and interpretation of typical experimental results (van der Poel *et al.*, 2005a). The propagation time of the light pulse can be controlled by the pulse itself (*self-induced speed control*) or by injection of a second pump pulse (*external speed control*). In an experimental configuration (van der Poel *et al.*, 2005a) the latter situation corresponds to a typical pump–probe setup. We here consider the propagation of a pulse with a duration of 180 fs. This represents a broad-band light signal. The space-time modelling allows to systematically vary the input pulse energy and injection current and calculate the resulting pulse shift. Figure 6.2 summarises the theoretical results. The pulse shift is considered relative to its position in the unsaturated device and calculated relative to the pulse duration.

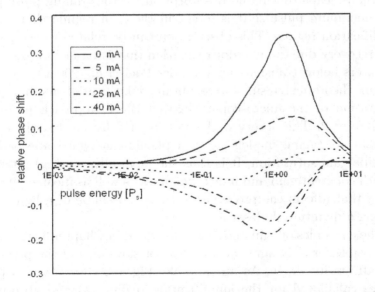

Fig. 6.2. Relative pulse shift in dependence on input pulse energy for different currents.

We would like to note that a logarithmic scale has been chosen for the input energy. The theory allows a prediction and interpretation of typical experimental results (van der Poel *et al.*, 2005b).

In the amplification regime (i.e. at injection currents larger than the transparency current of 15 mA) a gradual relative pulse acceleration of up to 20% is reached with increasing pulse energy. This corresponds to an absolute value of 36 fs. Near saturation, the leading edge of the propagating light pulse induces a dynamic level hole burning in the charge carrier system. As a consequence the gain available for the trailing part of the pulse is reduced. In combination with the pulse-induced index dispersion this leads to a pulse acceleration, i.e. a positive pulse shift. For pulse energies significantly beyond the saturation energy the pulse shift decreases. In this regime the level occupation is already entirely depleted by the very front part of the pulse.

In the absorption regime the sign of the relative pulse shift changes and relative pulse shifts of up to 40% are obtained (corresponding to an absolute value of 72 fs). Thereby, an increase in pulse energy leads to a bleaching of the absorption by the rising edge of the pulse. This eventually leads to a reduced absorption for the trailing pulse part. The maximum pulse shift is larger in the gain regime than in the amplification regime. This observation can be related to the partial gain recovery due to intra-dot relaxation that is typical for quantum dot lasers below transparency (van der Poel *et al.*, 2005a).

Our theoretical results were obtained by spatial and temporal integration of the microscopic Maxwell–Bloch approach presented in Chapter 2. The observed effects cannot fully be described with spatially averaging models using a phenomenological approach to describe gain saturation. It is thus the (spatially varying) interplay of both coherent gain and index dynamics as well as incoherent scattering that affects the propagation time and shape of a light pulse passing a quantum do laser amplifier.

The dynamics of gain and index discussed in Chapter 3 is not only responsible for a dynamic acceleration or slowdown of the pulse but also affects its shape. As an example, Fig. 6.3 shows typical pulse shapes calculated for the amplification regime (a) and absorption regime (b). The results agree very well with typical experimental

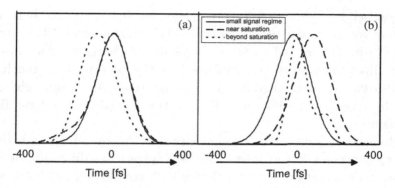

Fig. 6.3. Calculated pulse shapes after propagation in the quantum dot nanomaterial. (a) and (b) calculated pulse shapes for the amplification (a) and absorption (b) regime.

measurements (van der Poel *et al.*, 2005b). Figure 6.3 shows the pulse shape for input pulse energies in the small signal regime (solid line), near saturation (dashed line) and beyond the saturation regime (dotted line). The figures confirm the temporal shift of the pulse centre discussed in the previous section. In addition, a typical dynamical asymmetry can be seen near saturation. This asymmetry is particularly pronounced in the amplification regime. In both cases (i.e. amplification and absorption regime) the shift is a direct consequence of the dynamic depletion and generation of carriers induced by the light pulse during propagation. In addition, incoherent scattering processes such as, for example, intra-dot carrier relaxation via emission and absorption of phonons occurs. The influence of these incoherent processes thereby depends strongly on the energy difference between neighbouring dot levels (compared to the LO phonon energy). An energy difference near the LO phonon energy leads to a dynamic exchange of carriers (via emission and absorption of phonons) between the levels resulting in a rather uniform depletion of all levels. In a dot system with more decoupled levels a more selective depletion of individual levels (with highest dipole matrix elements and level energies in resonance with the pulse energy) will occur (van der Poel *et al.*, 2005a). As a consequence, saturation in a dot system with decoupled levels is already reached for smaller pulse energies. Another influence of the complex scattering dynamics occurs via the dipole dynamics that is (via the quantum Bloch equations) directly related to the changes in the charge carrier system. The dipole dynamics then affects (via the polarisation) the light fields dynamics described in the wave equations. As a result, changes in the level occupations are directly transferred to the light field dynamics.

In order to gain further insight into the influence of dynamic scattering processes on the pulse shape we contrast two different dot systems: (i) a quantum dot laser device with coupled energy levels and (ii) a quantum dot laser device with decoupled dot levels, i.e. where the dot levels are separated by more than the LO phonon energy so that emission and absorption of phonons is negligible. Figure 6.4(a) and (b) summarise the results of the simulations for the decoupled (a) and coupled (b) system. The figure displays the pulse shape evolution at the output facet in the absorption regime ($E = 0.5E_s$). The quantum dot ensemble with decoupled dot levels (Fig. 6.4(a)) leads to a comparatively smooth pulse envelope. In contrast, the shape of a light pulse after the passage in the dot ensemble with coupled energy levels (Fig. 6.4(b)) is characterised by irregular modulations. In this situation, the transfer of carrier scattering processes to the propagating light field dynamics induces a complex reshaping of the light pulses. This is particularly pronounced in the trailing part of the

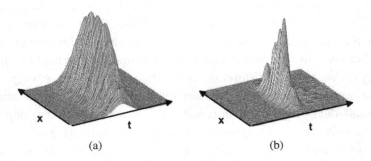

(a)   (b)

Fig. 6.4.   Spatially resolved pulse shapes at the output facet ($E = 0.5E_s$) in a quantum dot laser with decoupled dot levels (a) and in a quantum dot laser with energy levels coupled by dynamic carrier–phonon relaxation processes (b).

light pulse where a complex carrier relaxation occurs in response to the pulse excitation.

6.2.2 *Propagation control by a second pulse*

The complex dynamics of the polarisation induced by a propagating light pulse itself directly leads to the idea to control the propagation time with a second (pump) pulse (corresponding to a pump–probe experiment). In a next step, we will thus expose the active nano medium with a high-energy pump pulse before injecting a (probe) pulse. The energy of the pump pulse has been set near the saturation energy of the amplifier and the input energy of the probe pulse lies in the small signal regime. The duration of the two pulses is 170 fs and their wavelength is 1273 nm. This situation allows us to 'tune' the propagation time by varying two parameters: the delay between the two pulses and the injection current in the active quantum dot medium. Figure 6.5 shows the theoretically derived delay of the probe pulse in dependence on injection current and delay between pump and probe pulse. For the simulations visualised in Fig. 6.5(a) the delay between pump and probe pulse has been set to zero. At 0 mA a maximum relative delay of 12% is obtained (corresponding to an absolute delay of 20 fs). Increasing the current results in a decrease

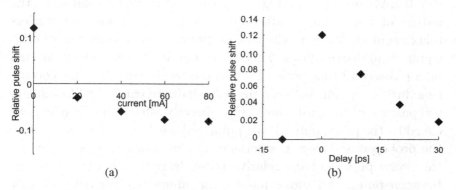

(a) (b)

Fig. 6.5. Pump-induced relative delay of a probe pulse in dependence on injection current and pump–probe delay: (a) dependence on injection current for zero pump–probe delay, (b) dependence on delay between pump and probe pulse at 0 mA.

in the delay and to a change in sign. A thorough adjustment of the current may then allow to 'tune' both the magnitude and sign of the propagation speed relative to the position of the probe pulse propagating (without presence of the second pulse) in the unsaturated medium. The results of a systematic variation of the delay are summarised (for zero injection current) in Fig. 6.5(b). The theoretical results confirm typical experimental measurements (Gehrig *et al.*, 2006). The physical processes causing the shifts are the same as in the situation of the single pulse propagation: the excitation of the charge carrier ensemble induced by the pump pulse leads to dynamic gain and index distributions. The observed dependence on the injection current (Fig. 6.5(a)) can be explained as follows: for low injection currents the probe pulse interacts with almost empty dots. This leads to a dynamic accumulation of carriers. The gain and index distributions corresponding to this carrier distribution then led to a delay of the probe pulse. For higher injection current levels the dots are significantly filled before their interaction with the pulses. As a consequence, a dynamic level hole burning arises leading to a speed-up of the propagating probe pulse. For a given value of the probe energy the injection current can thus be used to tune the relative shift experienced by the probe pulse.

The dependence of the propagation speed on pump–probe delay (for 0 mA) displayed in Fig. 6.5(b) can directly be related to the nature of the carrier scattering: after the excitation the carriers relax via intra-dot scattering and carrier capture towards their quasi-equilibrium distributions. This relaxation dynamics typically occurs on a picosecond time scale. The complex scattering dynamics occurring during the time between the pulse-induced carrier excitation and the passage of the probe pulse thus directly affects the delay experienced by the probe pulse. If the pump pulse is behind the probe pulse the probe pulse does not experience any modifications generated by the probe pulse and the relative pulse delay is zero. At zero delay between pump and probe pulse a maximum relative delay of 12% of the probe pulse could be obtained (corresponding to an absolute value of 34 fs). With increasing pump–probe delay the probe pulse experiences the gain dynamics and index dispersion of the partially relaxed carrier system leading to a characteristic decrease of the relative pulse shift.

Our simulation results demonstrate that the delay of the emitted light pulse strongly depends on injection current and input pulse energy and thus may be tuned by varying these parameters. They indicate the potential of active nano-structured media for a speed control of ultrashort, high bandwidth light pulses.

6.3 Conclusion

We have analysed the nonlinear interplay of propagating ultrashort (i.e. broadband) light pulses with active quantum-dot nanomaterials. Our results demonstrate a strong dependence of pulse shape and speed on injection current and input power, allowing to achieve controlled 'slow light' and 'fast light'. Furthermore, the properties of the emitted pulse are directly influenced by the energetic structure of the nanomaterial. The observed effects are shown to be related to the dynamics of the complex and ultrafast carrier dynamics discussed in Chapter 3. The dynamic carrier-induced spatio-spectral shaping of gain and index dispersion thus leads to measurable changes in relevant output parameters such as power, shape, duration and speed of a propagating signal. A fundamental knowledge of these processes may thus open new ways for a design and tuning of pulse properties.

References

Gehrig, E., Poel, M., Mørk, J., Hvam, J.M. and Hess, O., *IEEE J. Quantum Elect.* **42**, 1047–1054, (2006).

Ghosh, S., Pradhan S. and Bhattacharya, P., *Appl. Phys. Lett.* **81**, 3055–3057, (2002).

Hatori, N., Otsubo, K., Ishida, M., Akiyama, T., Nakata, Y., Ebe, H., Okumura, S., Yamamoto, T., Sugawara, M. and Arakawa, Y., Paper Th4.3.4, 30th European Conference on Optical Communication, (2004).

Hess, O. and Kuhn, T., *Prog. Quant. Electron.* **20**, 85–179, (1996).

van der Poel, M., Gehrig, E., Hess, O., Birkedal, D. and Hvam, J.M., *IEEE J. Quantum Elect.* **41**, 1115–1123, (2005a).

van der Poel, M., Mørk, J. and Hvam, J.M., *Opt. Express* **13**, 8032–8037, (2005b).

Chapter 7

High-Speed Dynamics

7.1 Mode-Locking in Multi-Section Quantum Dot Lasers

One of the advantages of quantum dot lasers is that the carrier dynamics in these nano-structured media occurs on ultrashort time scales. The resulting short carrier response allows a fast current modulation of the laser. Mode-locked laser systems are particularly attractive for high-speed telecommunication systems since they allow the generation of pulse sequences at repetition rates well beyond the modulation bandwidths of semiconductor lasers (Derickson et al., 1992; Kuntz et al., 2005a; Lau, 1990). Since most applications of mode-locked lasers require synchronisation with an external electrical signal it is thus common method to combine a passive pulse generating process realised by e.g. a saturable absorber with an active synchronising signal. This can be done by applying a radio-frequency signal to one of the laser sections, typically with a frequency in the vicinity of the cavity round trip frequency (Bimberg et al., 2006).

The description and modelling of mode-locked lasers is particular challenging since the physical processes that are important for the generation and shape of the pulses occur on various time scales (fs ... ns) and strongly depend on the geometry of the device. Thereby, the performance of the devices is strongly affected by the dynamic interplay of nonlinear mechanisms such as counter-propagation effects, spatio-spectral saturation, induced index and group velocity changes that may lead to timing jitter (Mulet and Moerk, 2004) and to pulse-to-pulse variations in amplitude and shape. In this chapter we will present selective results of numerical simulations of passive mode-locking in multi-section quantum dot lasers emitting at 1.3 μm. For a discussion and interpretation of the basic effects we will consider the two-section geometry (see Fig. 1.4 in Chapter 1). The device geometry consists of a short absorbing section and a long gain section responsible for pulse shaping and amplification. The simulations will be based on the multi-mode Maxwell–Bloch description (see Chapter 2) that describe the spatio-temporal light field and inter-/intra-level carrier dynamics in each quantum dot of a typical quantum dot ensemble (Gehrig and Hess, 2003). Experimental studies on mode-locked quantum dot lasers have demonstrated frequencies up to 50 GHz with pulse durations down to a few ps (Huang *et al.*, 2001; Kuntz *et al.*, 2005b; Thompson *et al.*, 2005). However, up to now the dependence of pulse duration on gain current and geometry of the two sections, which is vital for the application, has not been fully understood. Our model self-consistently includes all processes that are relevant in the mode-locking process. It thus represents a numerical tool that allows the investigation of the fundamental processes that limit the achievable pulse duration. In the following, we present theoretical results on passive mode locking of two-section lasers emitting at 1.3 μm. We demonstrate passively mode-locked pulses at repetition rates of 20, 40 and 80 GHz with a minimum duration of 1.5 ps. The pulse emission thereby strongly depends on bias voltage, injection current and geometry. We consider quantum dot systems consisting of fifteen InGaAs quantum dot layers with a stripe width of 4 μm. The total lengths of the quantum dot devices are 2000, 1000 and 500 μm corresponding to round trip frequencies of 20, 40 and 80 GHz, respectively. The short absorber section of our model system is operated at reverse bias levels between -3 and -7 V. The length of the absorber section is set to 1/10 of the total length.

7.2 Dependence of Pulse Duration on Injection Current, Bias Voltage and Device Geometry

In the model the two sections of the devices are represented by un-pumped absorbing material and inverted medium of variable carrier injection level respectively. A variable carrier life time is used to simulate a variable absorption in the short section corresponding to a variable reverse bias voltage. Details on the correlation between carrier life time and reverse bias current can be found in (Karin *et al.*, 1994; Williamson and Adams, 2004). After switching on the injection current the dynamic interplay of light diffraction and carrier scattering leads to a complex carrier-light field dynamics and to characteristic dynamic light patterns in the μm regime. After a few round trips in the cavity these spatially and temporally varying light field modulations then lead (via the intensity-dependent absorption in the short section of the quantum dot device) to the formation of light pulses. The integration within the framework of the wave equation thereby automatically leads to a self-consistent inclusion of the saturation characteristics of the absorber section, i.e. the absorption loss decreases with increasing intensity. In order to characterise the performance of the quantum dot laser devices we will in the following discuss the dependence of the pulse duration (full width at half maximum) on gain current and absorber voltage. The absorption is varied within a regime where mode-locking occurs (Kuntz *et al.*, 2004). In the simulation the full pulse shape is automatically calculated via the space-time integration of the light fields while it has to be artificially extracted in an experimental autocorrelation curve. The results are summarised in Fig. 7.1 for the 20 GHz, (a) 40 GHz (b) and 80 GHz (c) device (for three bias voltages). All curves show a similar dependence on reverse bias voltage and gain current: an increase in reverse bias leads to increased absorption within the waveguide allowing for the generation of shorter pulses. An increase in current, on the other hand, leads to an increase in pulse duration. This originates from the laser-internal carrier dynamics and group velocity dispersion. These processes gain more importance in the high-current regime. At very high currents this may lead to a continuous wave contribution in

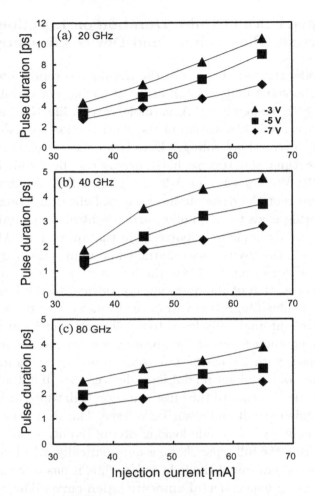

Fig. 7.1. Dependence of the pulse duration on gain current and reverse bias voltage for the passively mode-locked QD laser at (a) 20 GHz, (b) 40 GHz and (c) 80 GHz.

the laser emission, i.e. incomplete mode-locking: the interplay of carrier excitation and relaxation dynamics, amplified spontaneous emission (ASE) as well as light fluctuations leads to strong variations in spatio-spectral hole burning and index dispersion. This leads to dynamic modulations in the pulse shape, timing jitter, increased ASE between successive pulses and to a continuous wave contribution in the pulse emission (Haus, 1975; Mulet and Moerk, 2004). The shortest pulse durations are thus predicted for the regime of low gain currents and high reverse bias voltages. The quantum dot lasers with shorter length (i.e. larger frequency, see e.g. Fig. 7.1(c)) provide the emission of pulse sequences with smaller pulse durations than the devices with a larger cavity (e.g. Fig. 7.1(a)). In addition, the longer devices are characterised by a higher slope than the shorter lasers. This effect originates from the dynamic interplay between spontaneous and induced recombination: with increasing cavity length the spontaneous and induced emission processes between successive pulses become more pronounced. Their amplification in the large cavity leads to an increase in the light field dynamics. In the devices with the shorter cavities, the light fields experience less gain in the (smaller) spatial area in the time window between the pulses so that the influence on the current is reduced.

The influence of the length of the absorber section on the output characteristics (i.e. power and shape of the emitted pulse sequences) is not as large as that of the length of the gain section. However, our space-time simulations prove an influence of the length scales of the initial light fluctuations (typically in the μm regime) which enter the absorber section after switching on the laser. The spatial extension of the absorber consequently has to be sufficiently large in order to guarantee a spatio-temporal shaping of the light field that is the relevant effect responsible for the shortening of the pulses. We are thus able to assert that a minimum length of the absorber section of approximately 20 μm is recommended in order to guarantee a sufficient pulse shortening.

The influence of continuous wave emission processes is stronger in the 20 GHz device (i.e. the longest cavity) leading to partially incomplete mode-locking in the small current regime. This could also be observed in experiments (Kuntz *et al.*, 2007). We would like to note that the corresponding theoretical autocorrelation width (artificially simulated on the basis of the theoretical pulse durations) showed a very good agreement with experimental results (Bimberg *et al.*, 2006; Kuntz *et al.*, 2007).

7.3   Radio Frequency Spectra of the Emitted Light

In order to gain further information on the contributions in the laser emission we can perform a spectral analysis by applying a Fourier analysis to the emitted fields. The resulting spectra corresponds to the radio frequency (RF) spectrum that can be measured in a typical experimental setup. As an example Fig. 7.2 visualises the

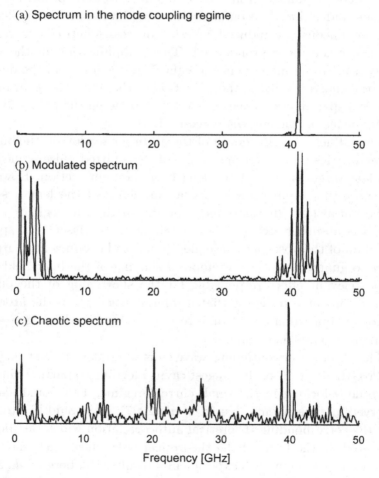

Fig. 7.2.   Calculated spectra in three typical mode coupling regimes: (a) pure mode locking, (b) modulated mode-locking and (c) chaotic behavior.

spectra for three characteristic regimes of the 40 GHz device laser: (a) pure mode-locking, (b) modulated mode-locking and (c) chaotic behaviour. These spectra are typical examples of regimes of operations discussed in literature (Bandelow *et al.*, 2006). Figure 7.2(a) shows a typical spectrum of pure mode-locking as obtained for sufficiently high absorption and intermediate current regime. In this case, the spectrum concentrates on a single-mode near 40 GHz. Please note that the vertical axis refers to a logarithmic scale. For small currents the interplay of spatial and spectral degrees of freedom leads to the coexistence of a number of modes (see also discussion above). This regime typically is characterised by modulated mode-coupling. In this case, the round trip in the resonator leads to the dynamic competition and coexistence of a frequency comb. In particular, a second mode group arises at the low frequency side of the spectrum. This behaviour is in agreement with typical experimental measurements (Kuntz *et al.*, 2007) (measurements with a fast photo detector and an electrical spectrum analyser). The modulations superimposed the 40 GHz peak are approximately 1 GHz corresponding to a time of 1 ns. This clearly indicates the influence of Q-switching. The corresponding contributions are generated around the mode-locking frequency and additionally around zero. The Q-switching frequency that is reflected in the frequency spacing increases with increasing gain current until a critical value for the transition to pure mode-locking is reached. Finally, the spectrum calculated for a high current (Fig. 7.2(c)), indicates a chaotic regime characterised by a large number of frequency contributions at irregular spacing.

The results discussed so far indicate that the occurrence of mode-locking is strongly dependent on injection current and bias voltage. A typical method to summarise these dependencies is the side mode suppression ratio (SMSR) with respect to the mode-locking frequency for varying gain currents and absorber voltages. Figure 7.3 summarises computational results as obtained for the 40 GHz device. In an experiment the SMSR is taken from the electrical power spectrum of an optical detector. In the simulation the SMSR is obtained as follows: firstly, a Fourier transformation is done with the time-dependent optical power obtained in the simulation. Secondly, the

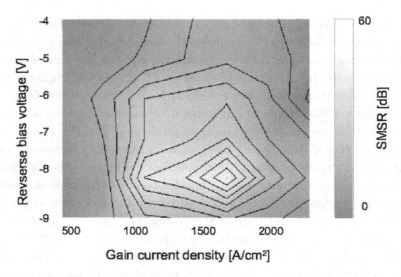

Fig. 7.3. SMSR plot: Displayed are the values relative to the mode-locking fundamental RF peak. Dark sections correspond to strong modulation of mode locking whereas grey and light shading denote pure mode locking with a SMSR > 20 dB.

amplitude ratio (in dB) between the carrier frequency (i.e. around 40 GHz at 1 mm length) and the next smallest side mode in the interval carrier frequency plus/minus half the carrier frequency (i.e. 40 ± 20 GHz) is taken from the spectral power density. The results displayed in the figure demonstrate that a regime of optimum mode-locking exists (i.e. high side mode suppression ratio). The sharp transition at the low-current side indicate the transition to modulated mode-coupling whereas the gradual decrease at the high-current side visualise the deterioration of the mode-coupling and the growing influence of the laser-internal dynamics.

7.4 Short-Pulse Optimisation

Using our advanced modelling tools we can additionally analyse the dependence of the pulse quality on the current. As an example, Fig. 7.4 shows the complexity of the underlying physical effects. Displayed are (in a time window of 0.14 ns) simulation results of the

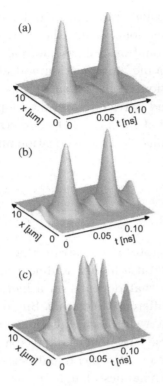

Fig. 7.4. Pulse sequency emitted by a 20 GHz quantum dot laser device for an injection current in the gain section of (a) 40 mA, (b) 60 mA and (c) 80 mA.

emission dynamics of the 20 GHz device. The plots refer to an injection current of (a) 40 mA, (b) 60 mA and (c) 80 mA in the gain section, respectively. At 40 mA the laser emits a regular pulse sequence with only marginal light emission between successive pulses. For the intermediate regime (60 mA, Fig. 7.4(b)) the simulation clearly demonstrates light field contribution between the pulses. Finally, for very high currents the coupling of the longitudinal modes is disturbed by the increased influence of the carrier dynamics leading (via the dynamic coupling of light and matter) to irregular pulsations and chaotic behaviour (Fig. 7.4(c)). The observed dynamics dependence of the pulse duration on current and bias voltage cannot be described on the basis of spatially averaged phenomenological models. Only the inclusion of the full spatial and temporal carrier and light field dynamics as done within the framework of the quantum dot Maxwell–Bloch equations allows the realistic simulation of the space-dependent dynamic amplification and absorption of the light fields in the two-section geometry as well as the mutual interplay of light diffraction and carrier scattering coupling the longitudinal and transverse degree of freedom. These processes are — for a given configuration — responsible for the duration and quality of the emitted pulses.

7.5 Conclusion

The results of this chapter demonstrate that nano-structured semiconductor devices exhibit a highly nonlinear and complex light field and carrier dynamics that depends on a hierarchy of physical processes occurring on different time scale. Simulations on the basis of the quantum dot Maxwell–Bloch equations (explicitly taking into account the transverse and longitudinal space coordinates) allow a realistic representation of typical spatio-spectral properties. Our results visualise the dynamics of light fields and carriers and provide insight into the coupling and interplay of spatial and temporal degrees of freedom. For a given laser structure and set of parameters the multi-mode Maxwell–Bloch equations allow efficient calculation of temporal emission characteristics and of passive mode-locking in two-section quantum dot lasers that can directly be compared with experiments. Spatio-temporally resolved calculations predict a strong dependence of the pulse duration on gain current and geometry. The increase of pulse duration with increasing current in the gain section could be related to the complex interplay of (spontaneous and induced) light emission processes and the charge carrier system in the dot ensemble. The coupled carrier and light field dynamics are of particular importance in long-cavity lasers where the high amplification of the light fields in the spatio-temporal regime between successive pulses enlarges the pulse duration and limits the mode-locking regime.

Our results show that a careful design of the quantum dot laser geometry is crucial for ultrashort pulse generation. Using the details of a given material system and laser geometry our space-time model thus opens the way to explore and predict the physical properties of state-of-the-art lasers and to contribute to the design and the development of optimised pulse sources.

References

Bandelow, U., Radziunas, M., Vladimirov, A., Hüttl, B. and Kaiser, R., *Opt. Quant. Electron.* **38**, 495–512, (2006).

Bimberg, D., Fiol, G., Kuntz, M., Meuer, C., Laemmlin, M., Ledentsov, N.N. and Kovsh, A.R., *Phys. Status Solidi A* **203**, 3523–3532, (2006).

Derickson, D.J., Helkey, R.J., Mar, A., Karin, J.R., Wasserbauer, J.G. and Bowers, J.E., *IEEE J. Quantum Elect.* **28**, 2186–2202, (1992).

Gehrig, E. and Hess, O., *Spatio-temporal dynamics and quantum fluctuations in semiconductor lasers.* Springer, Heidelberg (2003).

Haus, H.A., *IEEE J. Quantum Elect.* **11**, 736–746, (1975).

Huang, X., Stintz, A., Hua Li, Lester, L.E., Cheng, J. and Malloy, K.J., *Appl. Phys. Lett.* **78**, 2825–2827, (2001).

Karin, J.R., Helkey, R.J., Derickson, D.J., Nagarajan, R., Allin, D.S., Bowers, J.E. and Thorntonc, R.L., *Appl. Phys. Lett.* **64**, 676–678, (1994).

Kuntz, M., Fiol, G., Lämmlin, M., Bimberg, D., Thompson, M.G., Tan, K.T., Marinelli, C., Penty, R.V., White, I.H., Ustinov, V.M., Zhukov, A.E., Shernyakov, Yu. M. and Kovsh, A.R., *Appl. Phys. Lett.* **85**, 843–845, (2004).

Kuntz, M., Fiol, G., Lämmlin, M., Schubert, C., Kovsh, A.R., Jacob, A., Umbach, A. and Bimberg, D., *Electon Lett.* **41**, 244–245, (2005a).

Kuntz, M., Fiol, G., Lämmlin, M., Bimberg, D., Kovsh, A.R., Mikhrin, S.S., Kozhukhov, A.V., Ledentsov, N.N., Schubert, C., Ustinov, V.M., Zhukov, A.E., Shernyakov, Yu.M., Jacob, A. and Umbach, A., *Proc. of 13th Int. Symp. Nanostructures: Physics and Technology, St Petersburg, Russia*, p. 79, (2005b).

Kuntz, M., Fiol, G., Laemmlin, M., Meuer, C. and Bimberg, D., *P. IEEE* **95**, 1767–1776, (2007).

Lau, K.Y., *IEEE J. Quantum Elect.* **26**, 250–261, (1990).

Mulet, J. and Moerk, J., *Proc. SPIE* **5452**, 571–582, (2004).

Thompson, M.G., Marinelli, C., Zhao, X., Sellin, R.L., Penty, R.V., White, I.H., Kaiander, I.N., Bimberg, D., Kang, D.-J. and Blamire, M.G., *Electron. Lett.* **41**, 248–250, (2005).

Williamson, C.A. and Adams, M.J., *IEEE J. Quantum Elect.* **40**, 858–864, (2004).

Chapter 8

Quantum Dot Random Lasers

The complex carrier and dipole dynamics in semiconductor nanomaterials (as discussed in Chapter 3) reveals, particularly when incorporated in a laser, a vivid dynamic interplay of coherent and incoherent processes: on the one hand, the coherent coupling of light and matter leads to coherent laser emission. The complex carrier scattering and spatial inhomogeneity on the other hand imply incoherence. This inherent incoherence may affect both the spatio-spectral purity of the emitted radiation and the fidelity of a signal transmitted in an inverted nano medium.

This chapter focuses on the role and appearance of disorder and spatio-temporal coherence and gain in quantum dot ensembles. We will first introduce the general theoretical approach used to include the spatial inhomogeneity and disorder of an ensemble of semiconductor quantum dots together with the spatio-temporal dynamics. On this basis we then proceed to discuss the influence of homogeneous and inhomogeneous broadening on gain and on space-time coherence and conclude with a study of the dynamics of semiconductor quantum dot random lasers. Thereby, we fully include the dynamics of the (coupled system) and effects such as spatio-temporal gain saturation.

8.1 Spatially Inhomogeneous Semiconductor Quantum Dot Ensembles

Generally, the active semiconductor quantum dot gain material realised by self-organised epitaxial growth processes is characterised by a spatially inhomogeneous distribution of quantum dots (typically 10^6–10^7 for a device size of $10^4\,\mu\mathrm{m}^2$) that vary in size and position (Skolnik and Mowbray, 2004). In the following, we will outline a theoretical approach that explicitly includes the statistical properties of an ensemble of semiconductor quantum dots. For the numerical integration of Maxwell's wave equation we define a suitable numerical grid where the width and length of a mesh lie in the μm regime, respectively. In order to include the inhomogeneity of the dot medium within each mesh of this numerical grid we will in the following apply statistical methods.

To this end we introduce a representative subset of quantum dots that is formed by randomly choosing dots from an ensemble assuming that they follow a normal distribution. This will allow us to effectively model the physical properties of all quantum dots comprising the ensemble. The mean of the Gaussian distribution is given by the centre energy of the homogeneously broadened spectral line and the standard deviation is given by $\sigma = \Delta E_{FWHM} \cdot (2\sqrt{2\ln 2})^{-1}$ where ΔE_{FWHM} is the FWHM of the spectral line due to the inhomogeneously broadened QD ensemble. Parameters describing physical properties that may vary from dot to dot (like confinement energies, scattering rates and the coefficients describing spectral broadening and shift) are calculated and assigned for each representative QD before the start of the numerical integration. In this way inhomogeneous broadening — which is a characteristic property of an ensemble of quantum dots with varying sizes — is self-consistently included in the numerical model. Physical quantities that involve the whole dot ensemble are calculated by averaging over the total number N_{sim} of representative dots.

The gain spectrum of an inhomogeneously broadened QDSOA can be calculated with the following expression (Haug and Koch, 1998):

$$g(\omega) = \frac{\pi}{2}\frac{k_0^2}{\beta}\frac{\Gamma}{\epsilon_0\hbar}\frac{n_{QD}N_l}{d_a}\sum_{\kappa}\left[\sum_{i,j}(\Delta_{i,j}^{\kappa}\times\Delta_{i,j}^{\kappa*})\mathbf{e_x}\right.$$

$$\left.\times[n_i^{e,\kappa}+n_j^{h,\kappa}-1]L(\omega,\omega_{j,i}^{0,\kappa},\gamma_{j,i}^{\kappa})\right]N_{sim}^{-1}, \qquad (8.1)$$

where d_a is the width of the active area and N_l is the number of dot layers. The index κ runs over-all quantum dots in the statistical ensemble while i and j refer to the levels of electrons and holes, respectively. We assume s-polarisation in the lateral (i.e. x-) direction. The dipole matrix element tensor thus is multiplied with the unit vector in x-direction $\mathbf{e_x}$. Furthermore, we have used the following definitions: The homogenous line-width (transition $i \leftrightarrow j$) of quantum dot κ is

$$\gamma_{j,i}^{\kappa} = \gamma_{j,i} + n_{2d}^{e} A_{j,i}^{e,\kappa} + n_{2d}^{h} A_{j,i}^{h,\kappa}, \tag{8.2}$$

the central frequency (transition $i \leftrightarrow j$) of quantum dot κ is

$$\omega_{j,i}^{0,\kappa} = \omega_{j,i}^{e,h,\kappa} + n_{2D}^{e} B_{j,i}^{e,\kappa} + n_{2D}^{h} B_{j,i}^{h,\kappa}, \tag{8.3}$$

and the normalised Lorentzian line function centred at $\omega_{j,i}^{0,\kappa}$, is given by

$$L(\omega, \omega_{j,i}^{0,\kappa}, \gamma_{j,i}^{\kappa}) = \gamma_{j,i}^{\kappa}(2\pi[(\omega - \omega_{j,i}^{0,\kappa})^2 + (\gamma_{j,i}^{\kappa})^2])^{-1}. \tag{8.4}$$

8.1.1 *Gain spectra*

In the following we will apply Eq. (8.1) to calculate and compare gain spectra of homogeneously broadened and inhomogeneously broadened quantum dot semiconductor optical amplifiers (QDSOAs).

Figure 8.1 shows the gain spectrum of a single quantum dot and gain spectra obtained by successively averaging over an increasing number of inhomogeneously broadened dots. We have assumed typical values for the homogeneous (Lorentzian) broadening of 3 meV for the ground state and 5 meV for the transitions involving the excited states. In combination with the spectral broadening originating from elastic scattering with 2D carriers (see also Section 3) this leads to a FWHM of approximately 6.5 meV for the ground state transition. The gain spectra of the ensembles consisting of a variable number of QDs have been calculated assuming an inhomogeneous broadening of 35 meV for the ground state transition and 50 meV for transitions involving excited states (Bimberg *et al.*, 1999). Similar spectra have been obtained in experiments using photoluminescence spectroscopy (Schreiber *et al.*, 2007). In order to obtain comparable gain spectra (when averaged over the whole dot ensemble) for the homogeneously and the inhomogeneously broadened

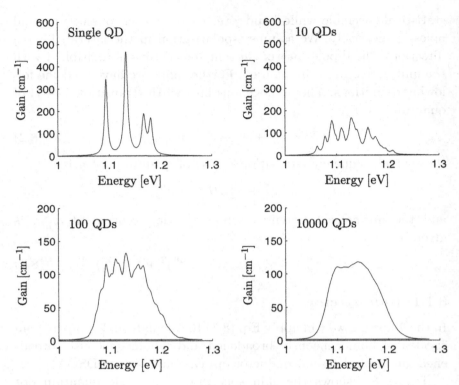

Fig. 8.1.   Gain spectra of a single quantum dot and a varying number of inhomogeneously broadenend QDs.

device, we have broadened the spectral line-width of the quantum dots in the homogeneously broadened QDSOA. Thereby, we have used a FWHM of 35 meV for the ground state transition and a FWHM of 40 meV for the transitions involving excited states. The corresponding gain spectra are summarised in Fig. 8.2. The results reveal a strong dependence of the QDSOA characteristics on spatial fluctuations of the dot properties.

8.1.2 *Spatial and spectral hole burning*

In a next step we will discuss the interaction of an injected optical pulse with the quantum dot ensemble within the active area of the QDSOA. Figure 8.3 shows snapshots of the gain spectra

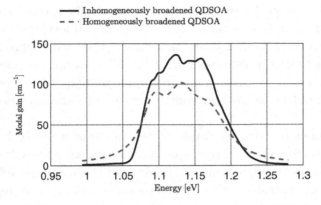

Fig. 8.2. Gain spectrum of an inhomogeneously broadenend QDSOA (FWHM GS: 30 meV, FWHM ES: 50 meV) and a homogeneously broadened QDSOA (FWHM GS: 25 meV, FWHM ES: 45 meV).

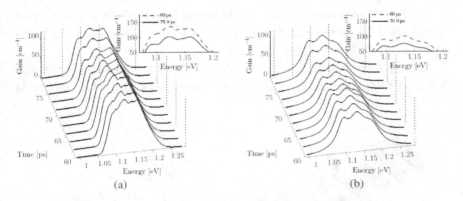

Fig. 8.3. Gain spectra of (a) an inhomogeneously broadened QDSOA and (b) a homogeneously broadenend QDSOA during the amplification of an optical pulse with a duration of 500 fs (injected at 64 ps).

of an inhomogeneously and a homogeneously broadened QDSOA, respectively, taken during the propagation the optical pulse along the waveguide cavity. In the case of the inhomogeneously broadened QDSOA (Fig. 8.3(a)) a pronounced hole burning can be seen at the energy of the injected optical pulse (1.1 eV). In contrast, the spectra of the homogeneously broadened QDSOA shows an over-all reduction of the gain caused by stimulated emission via the ground state transition and relaxation of carriers from the excited levels to the ground state: the ideal QDSOA contains identical QDs with a ground state transition that is in resonance with the injected optical pulse. Thus, every QD in the electrically pumped area participates in the process of stimulated emission and contributes to the amplification of the pulse.

The quantum dot ensemble of the inhomogeneously broadened QDSOA is characterised by a spatially dependent energy level structure of the dots. Thus, only a subgroup of the total ensemble has a ground state transition that is in resonance with the injected optical pulse. This subgroup strongly interacts with the optical pulse. This process is illustrated by the plots in Fig. 8.4 showing the spatially resolved modal gain (along a central axis of the waveguide parallel to the propagation direction) at the wavelength of the injected optical pulse. Note the striking difference between the spatially resolved gain of the ideal QDSOA (Fig. 8.4(a)) and the inhomogenously broadened QDSOA (Fig. 8.4(b)), respectively.

The hole burning induced by the probe pulse (injected from the left at 64 ps) during its propagation through the QDSOA cavity is

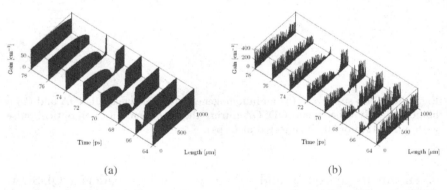

(a)   (b)

Fig. 8.4.   Spectral hole burning due to an injected optical pulse of duration 500 fs and population recovery of the electron ground state in (a) a homogeneously broadened QDSOA and (b) an inhomogeneously broadened QDSOA. The plots show the modal gain at the wavelength of the injected pulse along a central axis of the waveguide in propagation direction. The optical pulse is injected from the left at 64 ps.

clearly visible. Dots that are in resonance with the optical pulse are strongly depleted and contribute to the absorption of light in the active area before they are refilled within a few picoseconds.

8.2 Coherence Properties

Disorder within a quantum dot ensemble not only has a significant influence on the spatio-spectral gain but also on the coherence properties of the emitted light. In order to study the influence of both spatial disorder effects and spatio-temporal coherence, we have systematically varied the amount of homogeneous and inhomogeneous broadening of the quantum dots in a range between 1 meV and 30 meV. The space-time simulations were done for a period of 2 ns, respectively.

Generally, spatial disorder reflected by varying size and positions of the dots directly leads to changes in the space-time correlations of the light fields: firstly, the space-dependent level energies lead to changes in the dipole transitions of the dots. Those are, via the spatially propagating fields, directly transferred to corresponding changes in real and imaginary part of the light fields. Secondly, the inhomogeneous size and localisation of the dots leads to space-dependent carrier scattering processes that also affect the space-time coherence.

One method to analyse the space-time coherence is to monitor the cross-correlation of the light fields at the output facet of a quantum dot optical amplifier. This cross-correlation of the light fields may be defined via (Shiktorov, 1999)

$$C(x_0, \tau) = \frac{\langle \delta I(x_c - x_0) \delta I(x_c + x_0, t + \tau) \rangle_r}{\sqrt{\langle \delta I^2(x_c - x_0, t) \rangle_t} \sqrt{\langle \delta I^2(x_c + x_0, t) \rangle_t}}, \qquad (8.5)$$

where the bar denotes the time average.

To monitor the intensity distribution and the space-time correlations at the output facet of a quantum dot laser we will without loss of generality refer to a laser structure with a cavity length of 100 μm and a stripe width of 20 μm. The current density has been set to $5 \cdot 10^{-5}$ e/psnm^2. We will study, in particular, the influence of inhomogeneous (σ) and homogeneous line-width broadening (Γ) on the spatio-temporal dynamics and coherence properties of the emitted light.

In a first step of our study, we fix the homogeneous broadening and vary the inhomogeneous broadening, setting the homogeneous broadening to $\Gamma = 1$ meV and varying the inhomogeneous broadening σ is a regime between 10 and 30 meV. This variation is in agreement with typical values than can be found in literature (Grundmann, 2000; Sugawara *et al.*, 2000). In a real experimental situation, the lower value for the homogeneous broadening would correspond to the low temperature behaviour. At room temperature Γ reaches values in the regime $10 \ldots 20$ meV. Figure 8.5 displays the (spatially resolved) dynamics of the intensity at the output facet (top) and the

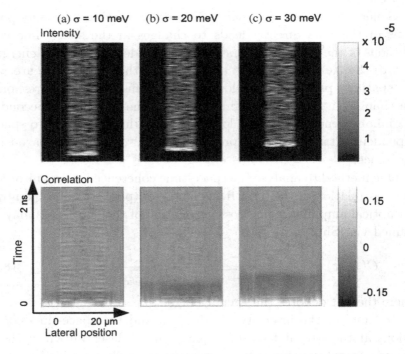

Fig. 8.5. Dynamics of the light fields (*top row*) and cross-correlations (*bottom row*) for a homogeneous broadening of $\Gamma = 1$ meV and an inhomogeneous broadening σ of (a) 10 meV, (b) 20 meV and (c) 30 meV.

corresponding correlation traces (bottom) in a time window of 2 ns. The distributions have been taken immediately after switching on the laser. The horizontal axis shows the dependence on the lateral coordinate at the output facet, the vertical axis denotes the time.

Generally, the large stripe width of 20 μm leads — for all values of the inhomogeneous broadening — to a complex transverse light field dynamics. In the case of high disorder the carrier density has to rise to higher values to start coherent lasing. As a consequence, an increase in inhomogeneous broadening leads to a delayed onset of the relaxation oscillations. Immediately after the start the dynamics of the correlations shows negative correlations (dark shading in Fig. 8.5) indicating the regime of the first laser relaxation. With increasing inhomogeneity, the duration of negative correlations gets longer. Furthermore, the dynamics in the distributions of the intensity and correlations changes with increasing disorder: for moderate inhomogeneous broadening the dynamics of the light fields and the correlations exhibit a rather uniform pattern. The transverse modes characterising the dynamics of the light fields are clearly resolved and reflected in the space-time correlations. An increase in the inhomogeneous broadening, however, leads to a spatial and temporal smearing-out of the distributions.

In a next step we analyse the influence of homogeneous broadening on the dynamics of light fields and space-time correlations. For that purpose, we set the inhomogeneous broadening to a comparatively small value, $\sigma = 1$ meV. In real semiconductor devices the inhomogeneity is much larger. However, setting the inhomogeneous broadening to this artificially low value allows us to isolate the influence of homogeneous broadening.

Figure 8.6 summarises for $\Gamma > \sigma$ the dependence of homogeneous broadening on intensity and space-time correlations. For the situation of low homogeneous and low inhomogeneous broadening (Fig. 8.6(a)) the intensity distribution is characterised by a clear pattern of distinguishable transverse modes. In a first time regime of approximately 1 ns the corresponding correlation plot shows strongly positive correlations near the stripe edges. After that regime the correlations at the edges change their sign towards negative values. This originates

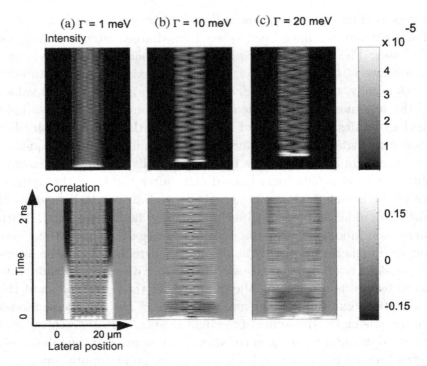

Fig. 8.6.   Dynamics of the intensity (*top row*) and cross-correlations (*bottom row*) for an inhomogeneous broadening of $\sigma = 1\,\mathrm{meV}$. The homogeneous broadening Γ has been set to (a) $\Gamma = 1\,\mathrm{meV}$, (b) $10\,\mathrm{meV}$ and (c) $20\,\mathrm{meV}$.

from the high density of carriers that accumulate in these areas. With increasing homogeneous broadening (Fig. 8.6(b) and (c)) the intensity distribution shows an increasing number of transverse modes (approximately two modes for $\Gamma = 10\,\mathrm{meV}$ and three modes for $\Gamma = 20\,\mathrm{meV}$). This is a direct consequence of the finite life time of the transitions: for larger homogeneous broadening an increased number of transverse modes fit in the spectral regime defined by the line-width. The various spectral contributions are also reflected in the correlation traces: the transverse zig-zag dynamics of the light fields leads to corresponding changes in sign in both spatial and temporal coordinates.

In a final step we will in the following discuss the influence of homogeneous and inhomogeneous broadening when the stripe width

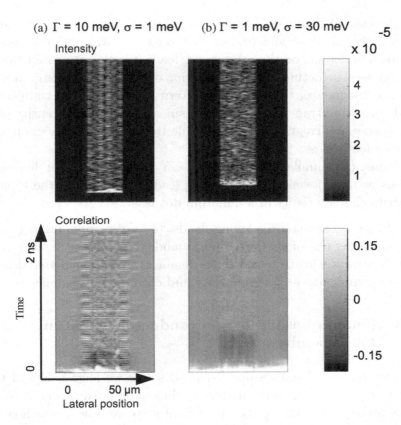

Fig. 8.7. Intensity (top) and correlations (bottom) for a stripe width of 50 μm. The broadening of the linewidth is dominated by (a) homogeneous broadening and (b) inhomogeneous broadening.

is enlarged. Figure 8.7 shows the dynamics of the intensity (top) and of the correlations (bottom) for (a) large homogeneous broadening and (b) large inhomogeneous broadening in a device where the stripe width has been set to 50 μm. In the case of homogeneous broadening (Fig. 8.7(a)) the homogeneous line-width is $\Gamma = 10$ meV and the inhomogeneous width has been set to a negligible value of 1 meV. The situation of dominant inhomogeneous broadening (Fig. 8.7(b)) refers to $\sigma = 30$ meV and $\Gamma = 1$ meV. The numerical results clearly reveal the role of disorder in space-time correlations: generally, the large stripe width leads to an increased number of transverse modes.

In the case of dominant homogeneous broadening these modes lead to a rather distinct and pronounced zig-zag movement. Inhomogeneous broadening, on the other hand, leads (via the space-dependent energy level structure) to the generation of new spectral components. As a consequence, the dynamic patterns formed by the competing and coexisting transverse modes are smeared out. The intensity plot thus shows an irregular structure while the correlation traces appear almost featureless.

Thus, our simulations show that an increase in either homogeneous or inhomogeneous broadening leads to a delay in the typical startup characteristics of a quantum dot laser:

(1) Inhomogeneous broadening leads to a dynamic mixing and smearing out of spatio-temporal mode dynamics.
(2) A variation in stripe width has demonstrated an increased degree of complexity in both intensity and correlation dynamics.

8.3 Random Lasing in Semiconductor Quantum Dot Ensembles

In the computational results discussed so far we have analysed the influence of disorder on the material gain and on the dynamic shaping of a propagating light pulse. In a final step we now study how to exploit spatial disorder to start a 'chaotic lasing' process.

8.3.1 *The physics of random lasing*

Random lasing has recently attracted intense attention: firstly, the phenomena of random lasing helps to understand coherent phenomena in disordered media on a fundamental level and, secondly, it can directly be applied in optoelectronic systems due to ease of preparation (i.e. no mirrors required) and due to the small size of typical random lasers down to few microns.

Lasing emerges from special random nano- and micro-cavities formed within the active medium (Burin *et al.*, 2003). Numerical studies on the basis of spatially and temporally resolved simulations allow us to study the performance and properties of such special lasers without the restriction to a mode-expansion in terms of the modes of the nano-cavities that is required in analytical approaches. It is easy to understand that since the pioneering work of Letokhov (1968) lasing in random media has been in the focus of a considerable number of studies (Beenakker, 1998; Burin *et al.*, 2003; Cao and Zhao, 1999; Cao and Xu, 2000; Cao, 2003; Cao *et al.*, 2003; Gaizauskas and Feller, 2003; Hackenbroich, 2005; Jiang and Soukoulis, 2000; Lawandy *et al.*, 1994; Liu *et al.*, 2004, 2005, 2006; Vanneste and Sebbah, 2001; van Soest *et al.*, 2001; Wiersma and Cavalieri, 2001; Zacharakis *et al.*, 2002).

In an active random medium, light is scattered and undergoes a random walk before leaving the device. Generally, one can differentiate between two kinds of random lasers: lasers with nonresonant (incoherent) feedback and lasers with resonant (coherent) feedback. Recent work has also considered the role of multi-mode emission on the basis of a semiclassical laser theory (Hackenbroich, 2005). In the case of incoherent feedback the scattering of light in the region around the initial light creation process leads to light amplification or, spoken in terms of the photon picture, to the generation of successive photons. The situation is different if light is amplified in resonance. This case occurs if light returns to the scattering centre forming a closed-loop path for light. In a spatially extended active nano-medium incoherent scattering is caused by carrier scattering and dipole dephasing whereas coherent contributions arise from either resonant amplification in resonant dots or from the geometrical feedback realised by the boundaries of the cavity.

In the quantum dot laser randomness is caused by spatial disorder in quantum dot size and positioning leading to space-dependent charge carrier and inter-level polarisation dynamics. Similar to (Hackenbroich, 2005) the random scattering thus enters the theory through the statistical properties of the decay rates and matrix elements which we assume (but are not limited to) being Gaussian distributed. The light that is locally created by stimulated emission may then travel to surrounding dots where a partially coherent amplification occurs. The coherence properties of the light in the neighbouring regions thereby depends on the coherent but spectrally detuned induced emission in the surrounding dots (which may be of different size and exhibit a different transition energy) and on incoherent processes such as carrier and dipole dephasing. The feedback provided by the disorder-induced scattering may thus have a stronger influence on the light field and carrier dynamics than the feedback realised by the laser cavity leading to interesting effects and spatio-spectral emission characteristics. This situation will be discussed in detail in this chapter.

The spatial inhomogeneity in the dot properties of a quantum dot active medium leads to the formation of nano- and micro-scaled cavities formed by the partial light scattering in the area surrounding a dot. The gain and loss of the individual nano- and micro-cavities thereby is determined by the difference in dot properties. In the following we will consider the two limiting cases of high and low disorder. Without loss of generality the percentage of the fluctuations will be set to identical amplitude in all material properties (scattering rates, dipole transitions). We would like to note, however, that the parameters may principally be varied and studied individually. The values for the fluctuations in the case of high disorder will be chosen such that they dominate — according to the above differentiation between coherent and incoherent feedback — over the influence of the geometrical feedback realised by the cavity. In the situation of low disorder the values will be comparatively small (typically less or equal to two percent) so that coherent feedback of the resonator dominates the behaviour. As we will see in the following this transition between high and low disorder strongly depends (as a natural consequence of the definition) on the absolute value of the facet reflectivity.

8.3.2 *Lasers with strong disorder: incoherent feedback*

A strong inhomogeneity in the material properties of the nanomaterial leads to a large spatial variation in the carrier and polarisation dynamics. In this case the reflectivities realised by the spatially varying dot properties are dominant and the reflectivity of the laser

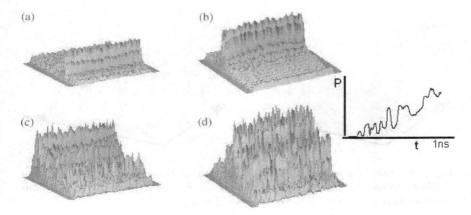

Fig. 8.8. Intensity snapshots ($\Delta t = 5\,\text{ps}$) in an optically injected (with a Gaussian shaped pulse of duration $1\,\text{ps}$) quantum dot semiconductor optical amplifier of high disorder. The facet reflectivity is $R = 10^{-4}$ and the input power is $P = 0.5P_s$ (where P_s denotes the power required to saturate the inversion). Bright shading indicates high intensity values.

cavity plays a minor role. The dynamics of light and matter thus is mostly affected by the incoherent feedback in the spatial dot array. Figure 8.8 shows a snapshot ($\Delta t = 5\,\text{ps}$) of the intensity distribution in an inverted quantum dot medium into which a coherent light pulse has been injected. The injected light pulse is Gaussian shaped (in lateral direction) and has a duration of 1 ps. In the shown example the input power is high ($0.5P_s$ where P_s denotes the power required to saturate the inversion in the medium) so that the carrier inversion is strongly reduced by the incoming light field. The propagating pulse light leads to a spatial excitation of the dots. The locally generated light fields are then strongly amplified by multi-reflection processes in the environment of the spatial nano-cavities. As a consequence, a dynamic increase in the over-all intensity of the devices arises leading to the built-up of lasing. Due to the small size of the nano-cavities this built-up is fully established within a few hundred ps (see plot on the right side of Fig. 8.8) and the facet reflectivities (in the given example: $R = 10^{-4}$) play a minor role.

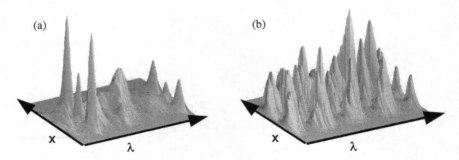

Fig. 8.9. Spatially resolved emission spectra of the optically injected nano medium discussed in Fig. 8.8. (a) Temporal average for the first 400 ps, (b) temporal average in the time window between 2 and 2.4 ns.

The spatial disorder of the medium not only affects the spatio-temporal propagation of light but also affects the spectral properties of the medium: in particular, the spatially varying dot size leads to space-dependent transition energies. As a consequence more spatio-spectral modes may arise leading to a complex spectrum. Figure 8.9 shows the corresponding emission spectrum of the discussed system. The spatially resolved spectrum is calculated at the output facet, temporally averaged for the first 400 ps (a) and in a time window between 2 and 2.4 ns (b). The spectra show a multitude of longitudinal and transverse modes that is typical for large area semiconductor lasers. For the first time window (Fig. 8.9(a)), however, the spatio-spectral properties of the intense input pulse leads to selection of modes, in spite of the large inhomogeneity of the system. After the passage of the pulse (Fig. 8.9(b)) the influence of the light injection is reduced and the laser is characterised by a multi-mode emission.

A similar behaviour is observed if no light is injected (Figs. 8.10 and 8.11). In this case the light that is locally generated by spontaneous emission is sufficient to start the multi-mode lasing behaviour. The snapshots displayed in Fig. 8.10 thus show a starting of a lasing behaviour that is very similar to the behaviour discussed in Fig. 8.8. Due to the completely self-induced light amplification the lasing starts in multi-mode emission (Fig. 8.11) without any additional coupling that would be established by a coherent injection (Fig. 8.8).

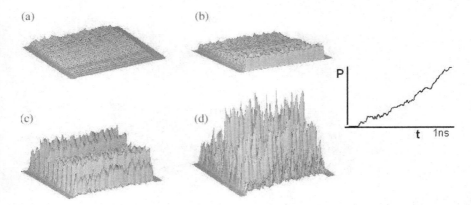

Fig. 8.10.   Intensity snapshots ($\Delta t = 5\,\text{ps}$) in a quantum dot semiconductor optical amplifier without optical injection and high disorder.

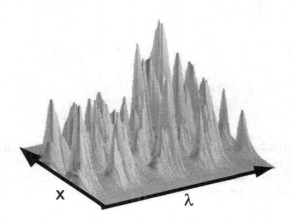

Fig. 8.11.   Spatially resolved emission spectra of the optically injected nano medium discussed in Fig. 8.10.

8.3.3 *Lasers with weak disorder: coherent feedback*

The situation is changed if we consider a medium with small spatial variance in the dot properties. In this case the dynamics of the system is to a higher degree determined by the facet reflectivities of the nano medium. As a consequence a minimum reflectivity of the facets and a non-vanishing input power is required to start the random lasing.

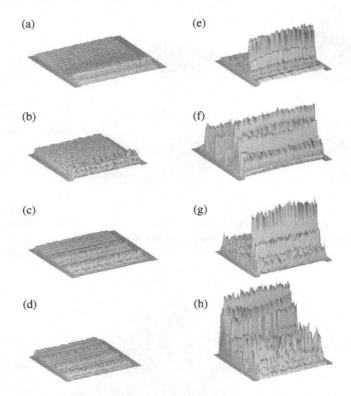

Fig. 8.12.   Intensity snapshots ($\Delta t = 5$ ps) in an optically injected (with a Gaussian shaped pulse of duration 1 ps) quantum dot semiconductor optical amplifier. The input power is 0.1 (*left*) and 0.5 (*right*) P_s (where P_s denotes the power required to saturate the inversion). The facet reflectivity is $R = 10^{-3}$.

Figure 8.12 shows snapshots of the intensity distribution in a semiconductor optical amplifier with a facet reflectivity of 10^{-3} and a spatial fluctuation in the material properties of 2 percentage. The input power has been set to 0.1 (*left*) and 0.5 (*right*) P_s, respectively. The figures clearly reveal the relevance of the input pulse for an initiation of the lasing process. The moderate (compared to the situation of high disorder) multi-reflection processes of the light fields in the immediate environment lead to the requirement of an additional initiating process for the local light amplification. This is provided by the input pulse. As a consequence, a minimum input power is required in order to supply a sufficient energy. In the simulations, we have systematically varied both facet reflectivites and input power values. As a general tendency we can summarise that the role of the facet reflectivities and of the input power somehow complement each other: for higher facet reflectivities low input power values are required to initiate the random lasing (e.g. $0.05P_s$ for $R = 0.02$); for smaller facet reflectivities comparatively small amplitudes are reached within the cavity (due to the higher outcoupling rate). As a consequence, a higher input power is required to reach the threshold for the random lasing process. Examples are input power values of $0.7P_s$ for a facet reflectivites of $R = 10^{-4}$.

8.4 Conclusion

In this chapter we have analysed and discussed the influence of spatial inhomogeneity and disorder on the physical properties of an ensemble of semiconductor quantum dots. Computational results reveal a strong dependence of the (spatio-temporally varying) gain on the characteristic spatial inhomogeneity of an ensemble of self-organised semiconductor quantum dots. We have shown that the spatial disorder that is typical for quantum dot structures leads to the formation of characteristic micro- and nano-cavities that may lead to the built-up of a random lasing process. Numerical simulations on the basis of spatially resolved Maxwell–Bloch equations reveal the dynamic interplay and the respective influence of facet reflectivities, input power, spatio-temporal saturation (e.g. gain saturation) and disorder in the lasing process. This builds the foundation for future studies aimed at an investigation of the photon statistics for a further fundamental analysis of the interplay of coherent and incoherent spatio-temporal light contributions.

References

Beenakker, C.W.J., *Phys. Rev. Lett.* **81**, 1829–1832, (1998).
Bimberg, D., Grundmann, M. and Ledentsov, N.N., *Quantum Dot Heterostructures*, John Wiley Sons, New York, (1999).

Burin, L., Cao, H. and Ratner, M.A., *Physica B* **338**, 212–214, (2003).

Cao, H. and Zhao, Y.G., *Appl. Phys. Lett.* **75**, 1213–1215, (1999).

Cao, H. and Xu, J.Y., *Appl. Phys. Lett.* **76**, 2997–2999, (2000).

Cao, H., *Wave Random Media* **13**, R1–R39, (2003).

Cao, H., Ling, Y., Xu, J.Y., Burin, A.L. and Chang, R.P.H., *Physica B* **338**, 215–218, (2003).

Gaizauskas, E. and Feller, K.-H., *J. Lumin.* **102**, 13–20, (2003).

Grundmann, M., *Appl. Phys. Lett.* **77**, 1428–1431, (2000).

Hackenbroich, G., *J. Phys. A* **38**, 10537–10543, (2005).

Haug, H. and Koch, S.W., *Quantum Theory of the Optical and Electronic Properties of Semiconductors*. World Scientific, Singapore (1998).

Jiang, X. and Soukoulis, C.M., *Phys. Rev. Lett.* **85**, 70–73, (2000).

Lawandy, N.M., Balachandran, R.M., Gomes, A.S.L. and Sauvain, E., *Nature* **368**, 436–438, (1994).

Letokhov, V.S., *Sov. Phys. JETP* **26**, 835–840, (1968).

Liu, C., Liu, J., Zhang, J. and Dou, K., *Opt. Commun.* **244**, 299–303, (2005).

Liu, Y.J., Sun, X.W., Elim, H.I. and Ji, W., *Appl. Phys. Lett.* **89**, 11111–11113, (2006).

Liu, X., Yamilov, A., Wu, X., Zheng, J.-G., Cao, H. and Chang, R.P.H., *Chem. Mater.* **16**, 5414–5419, (2004).

Schreiber, M., Schmidt, T., Worschech, L., Forchel, A., Bacher, G., Passow, T. and Hommel, D., *Nat. Phys.* **3**, 106–110, (2007).

Shiktorov, P., *Appl. Phys. Lett.* **74**, 723–726, (1999).

Skolnik, M. and Mowbray, D., *Ann. Rev. Mater. Res.* **34**, 181–218, (2004).

Sugawara, M., Mukai, K., Nakata, Y., Ishikawa, H. and Sakamoto, A., *Phys. Rev. B* **61**, 7595–7603, (2000).

Uskov, A.V., Nishi, K. and Lang, R., *Appl. Phys. Lett.* **74**, 3081–3083, (1999).

van Soest, G., Poelwijk, F.J., Sprik, R. and Lagendijk, A., *Phys. Rev. Lett.* **86**, 1522–1525, (2001).

Vanneste, C. and Sebbah, P., *Phys. Rev. Lett.* **87**, 183903–183906, (2001).

Wiersma, D.S. and Cavalieri, S., *Nature* **414**, 708–709, (2001).

Zacharakis, G., Papadogiannis, N.A. and Papazoglou, T.G., *Appl. Phys. Lett.* **81**, 2511–2513, (2002).

Chapter 9

Coherence Properties of Quantum Dot Micro-Cavity Lasers

Spatio-temporal coherence in active semiconductor nano-structures is of high importance for a detailed understanding of the fundamental light–matter coupling. Furthermore, it represents a key property to applications in telecommunication and signal processing. In this chapter, we explore the quantum memory of spatially inhomogeneous quantum dot ensembles that have been optically injected by a short signal pulse. We present a fundamental analysis of the dynamics of the field–field and field–dipole correlations that can be obtained from a spatially resolved modelling on the basis of quantum luminescence equations. These correlations directly reflect the role of quantum effects that are responsible for memory effects in such a system with a high coupling of spatial and temporal degrees of freedom. Space-time simulations reveal the role of spatial inhomogeneity, spectral detuning of the signal pulse and input power for applications in the storage or spatial transfer of quantum signals in novel memory devices.

9.1 Introduction

Light is a promising candidate for carrying information in both classical and quantum communication systems. Up to now, various experiments have focused on an exploration of the coupling of light and matter for memory effects. The realisation of a quantum memory systems principally requires precise knowledge of the coherence properties of the material under investigation. Typical quantum memory elements are based on atom and ion ensembles. Various investigations concentrated on atomic vapors or atomic ensembles (Black *et al.*, 2005; Chou *et al.*, 2004; Eisaman *et al.*, 2004; Julsgaard *et al.*, 2004; Kuzmich, 2003). The transfer of a quantum state between matter and light was demonstrated in (Matsukevich and Kuzmich, 2004). These systems can easily be described but need a large experimental setup. The integration of quantum dot media in novel memory elements may lead to a higher practicability for industrial utilisation. Quantum dots have narrow spectral linewidths and rapid radiative decay rates. Furthermore, they can easily be integrated into larger structures, e.g. arrays or micro cavities to improve efficiency. They are attractive for quantum experiments, e.g. for single photon sources (Santori *et al.*, 2002). The ability to deterministically transform quantum states of light to quantum states of matter is not only of high relevance for a study of very fundamental interactions but also represents a key tool in the development of light-based quantum technologies. Quantum dots may in the future become key elements in the engineering of nanomaterials for quantum memories. The investigation of the memory of nano-media may thus open the way for long distance quantum communication links or network-model quantum computers.

In this chapter we use quantum luminescence equations for quantum dot nanomaterials that can be derived from a fully quantum mechanical approach. Our approach represents a fully microscopic description that allows the calculation of carrier and field dynamics. It includes all relevant optical and electronical material properties and does not require any fit parameters that have to be considered in more phenomenological approaches. Microscopic carrier effects such as carrier–carrier and carrier–phonon scattering are explicitly taken into account. Using a spatially resolved description our model thus bridges the gap between spatially averaged microscopic quantum approaches and phenomenological models. The theory takes into account, in particular, field–field correlations and field–dipole correlations representing the origin for coherence effects. The description allows to obtain information on spatio-temporal coherence which is the key property for quantum memory.

9.2 Radial Signal Propagation and Coherence Trapping

A particularly interesting situation for future applications in quantum information processing is the interaction of an ultrashort light pulse with an active nanomaterial. We focus on a model system consisting of a quantum dot layer embedded in a vertical cavity that has been optically excited by a short (500 fs) pulse as schematically depicted in Fig. 9.1. The quantum dot ensemble (material system InGaAs) is organised in a two-dimensional layer of $10\,\mu$m $\times$ $10\,\mu$m that is optically excited by a light signal of $5\,\mu$m diameter — the carrier of quantum information.

As we will see, this configuration may lead to wave-like radial signal propagation or to the trapping of coherence. The pulse duration is small enough to induce an ultrashort response of the medium. At the same time, it is sufficiently long to be resolved in the spatial and temporal domain within a reasonable observation time. In future investigations we plan to concentrate on even shorter pulses with a duration that corresponds to a single oscillation of the light field (i.e. single photon pulses) representing a qubit.

The theory is based on quantum luminescence equations that describe the interaction of the (quantum) light field with the

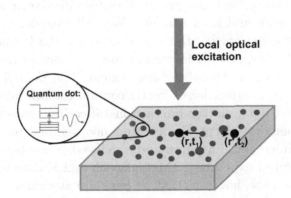

Fig. 9.1.   Scheme of the optical excitation of the quantum dot system in vertical-cavity configuration. $(\mathbf{r}, t_1)$ and $(\mathbf{r}', t_2)$ visualise the storage of information in r and the dynamic transverse transfer of information $(\mathbf{r} \to \mathbf{r}')$.

quantum states of the electrons and the holes in the semiconductor quantum dots. This approach includes on a microscopic basis the interactions and correlations of the light fields and the dipoles excited within the quantum dot ensemble that cause and mediate the (spatial and temporal) coherence of spontaneous and stimulated emission. It is summarised in Chapter 2. Using this approach we can analyse the spatial coherence properties relevant for quantum memory.

Relating a state at point $\mathbf{r}$ in the centre of the excited area to another state near the rim ($\mathbf{r}'$) gives a measure for the data stored or transferred after a given time. The injection of a light pulse (here: Gaussian shaped spatial beam profile, width $5\,\mu m$) high above the band gap into a semiconductor laser generates a (partially incoherent) excited electron-hole plasma which then leads to a hierarchy of carrier relaxation processes within the bands followed by (low-momentum) radiative carrier recombination that can be observed as (photo-) luminescence. The coherence properties of the emitted radiation are then directly reflected in the dynamics of the field–field correlation ($I(\mathbf{r}';\mathbf{r}, k_e, k_h)$) as well as of the dipole–dipole correlation ($C(\mathbf{r}';\mathbf{r}, k_e, k_h)$). Details of the theoretical approach can be found in Chapter 2. We use typical InGaAs material parameters (Gehrig and Hess, 2003). The respective epitaxial structure is taken into account via corresponding material parameters for effective masses, dipole matrix elements and level energies. We will consider resonant and off-resonant injection, i.e. the situation where the frequency of the injected pulse is spectrally located at the maximum of the gain curve or detuned towards the absorbing regime. The field–field (I) and field–dipole (C) correlations directly contain the full information on all correlated light–matter interactions affecting space-time coherence and memory effects. We are thus not required to additionally model the injection of a pump and a delayed probe beam as necessary in a typical experiment (Matsukevich and Kuzmich, 2004). We would like to note, however, that we here aim at a fundamental analysis of the requirements and conditions for efficient quantum systems

(a) Carrier density N^{eh}

(b) Field–field correlation I^{eh}

(c) Field–dipole correlation C^{eh}

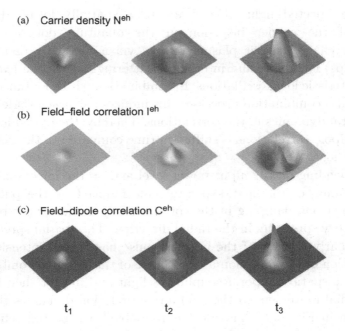

t_1 t_2 t_3

Fig. 9.2. Snapshots of the carrier density (a), field–field correlation (b) and field–dipole correlation (c) in the optically excited quantum dot ensemble.

based on light–matter coupling. The realisation of a long-time quantum storage, for example, would require a feedback loop conditioned on the signal transmitted by the system (Julsgaard *et al.*, 2004).

Figure 9.2 shows snapshots of the carrier distribution (a), field–field correlation (b) and field–dipole correlation (c) in the quantum dot layer optically excited by a resonant pulse. The maximum of the injected pulse (duration 500 fs) enters the active area in the first time step (left column). The time between successsive plots is 1 ps. The initial level occupation of the dots has been set slightly above transparency. In our example we chose a pulse with a high intensity so that long-lived responses can be expected in the carrier ensemble. The plots refer to the spectrally integrated values at the output facet. For each grid point the spatial correlations of all points have been summed up, i.e. $\tilde{I}^{eh} = \int_{\mathbf{r}'} \sum_{k_e, k_h} I(r'; r, k_e, k_h) dr'$. We then plot the real quantity $I^{eh} = \mathrm{Re}\,(\tilde{I}^{eh})^2 + \mathrm{Im}(\tilde{I}^{eh})^2$.

The injected light pulse leads to a spatially localised depletion of the carrier inversion in the quantum dot layer. The
excited charge carrier plasma relaxes via a hierarchy of ultrafast
(fs ... ps) intraband and interband scattering mechanism leading to
characteristic local oscillations. In combination with spontaneous and
induced recombination processes this induces a characteristic spatio-
temporal dynamics in the correlations. Thereby, the dynamics of the
field–dipole correlations is shifted in time compared to the field–field
correlations.

Depending on the input power level and on the injection density
the refilling of the spatio-spectral hole induced by the pulse may
lead to strong damping in the correlation dynamics or excite a fast
oscillatory spreading in the radial direction. The spatial spreading is
particularly efficient if the injected pulse has a high intensity. The
saturation of the inversion at the centre of the probe in combination
with the spatio-temporal coupling of light and matter then induces
the radial movement in the field and correlation dynamics that can
be seen in Fig. 9.2. This effect is reinforced if a partial refilling of
the levels in the centre is realised by a high carrier injection level.
The resulting changes in the carrier distribution keeps the changes
in radial direction at a high level leading to an additional pushing of
the radial dynamics.

A common method to quantify the quality of a quantum com-
puting device is to define a figure of merit (FOM) (Fiurasek, 2006;
Gardiner and Zoller, 2000; Matsukevich and Kuzmich, 2004). Here we
will, for practical reasons, define (in analogy to the FOM definition)
a correlation figure of merit for the field–field and field–dipole correlation at the centre and at the rim of the active area. For this purpose
we integrate the correlations to all other spatial grid points and nor-
malise them to their initial values. This directly reveals the quantum
memory storage time (Chaneliere *et al.*, 2005) of the QD ensemble as
well as the dynamics of coherence transfer. In other words, the FOM
provides information on the (time-dependent) quality of the storage
or the transfer of a coherent signal (here: the mixed light–matter
state induced by the light pulse).

Figure 9.3 summarises the dynamics of the field–dipole FOM
(a) and the field–field FOM (b) for resonant excitation. The left

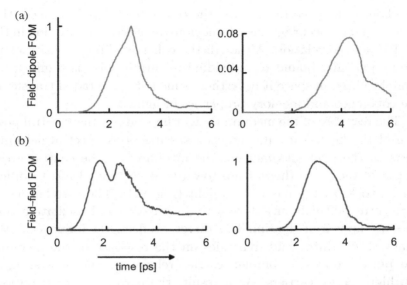

Fig. 9.3. Transfer of information. Field–dipole (a) and field–field (b) FOM in the centre (*left*) and at the rim (*right*) of the quantum dot medium: resonant injection.

column refers to the centre of the probe whereas the right column displays the dynamics of the correlations relative to a point at the rim (i.e. r'). The field–dipole FOM (Fig. 9.3(a)) is characterised by a significant decrease in magnitude. This orignates from the fact that the localised excitatation of the dipole is followed by a local decay of the dipoles. Nevertheless, approximately 10 % of the correlation has been translated to the rim (via the mutual inter-play of light and matter during the lateral expansion of the fields) just while they are decaying in the centre. The field–field FOM (Fig. 9.3(b)) is characterised by a slight reshaping of the curve. The second pulse that can be seen in the central distribution (left column) is a 'shadow pulse' arising from the strong excitation of the medium: the pattern imprinted in the medium by the excitation pulse induces the light–dipole coupling (Fig. 9.3(a)) which then feeds back to the light–light coupling. This effect is a direct consequence of the coupling of the correlations. Furthermore, the light–light FOM at the rim is of similar magnitude as in the centre. This directly demonstrates the fact that the dynamic field correlations are free to advance in space (here: from the centre to the rim). Indeed, this effect can only occur because of the non-vanishing contribution in the field–dipole correlation. Although the coherent dipole density alone (being partially bound to individual quantum dots) is significantly more localised in space it nevertheless mediates the radial transfer of the coherence and memory properties of light.

The response of the medium to light depends on the spectral position of the light field and the excited carriers. A spectral detuning between the gain maximum of the medium and the central wavelength of the pulse (here: 6 nm towards absorption) leads to different response of the medium to light (Fig. 9.4). The correlations at the centre (left column) show a characteristic local oscillation that does not appear at the rim. This directly reflects the influence of the quantum dot intraband dynamics on the response of the medium: the pulse induces a complex carrier relaxation and scattering of the high-energy carriers. As a result, the correlation distributions show characteristic modulations and the light–light FOM at the rim

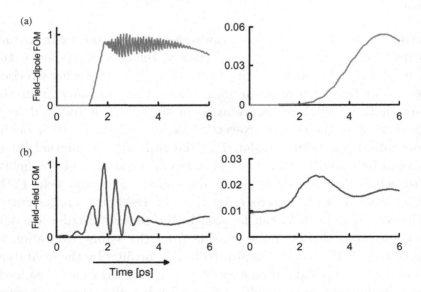

Fig. 9.4.   Storage of information. Field–dipole (a) and field–field (b) FOM in the centre (*left*) and at the rim (*right*) of the quantum dot medium: off-resonant injection.

(right column in Fig. 9.4(b)) is significantly decreased as compared to the initial value near the centre (left column in Fig. 9.4(b)). An off-resonant injection thus impedes the formation of the wave-like expansion of coherence and may lead to a spatial trapping of information. A spectral detuning may thus be a means of controlling the memory effects.

The influence of the spectral detuning can be further inferred from lateral cuts of the correlation functions. Figure 9.5 compares the dynamics of the real part of the field–field correlation function and the imaginary part of the diagonal elements of the field–dipole correlation function for a detuning of 6 nm (a) and 3 nm (b) to resonant excitation (c).

A strong detuning (Fig. 9.5(a)) leads to slight dynamic modulations on a ps time scale in the space-time correlations. These coherence oscillations are a direct footprint of the dynamic spatio-temporal interplay of fields and dipoles and do not originate from cavity round-trip effects. An off-resonant light field leads to a delayed response of the medium, i.e. the light–dipole correlations increase only after the first maximum of the field–field correlation. The induced changes in the dipole density, on the other hand, have a delayed influence on the light fields leading to a second maximum in the field–field correlation. As a result, the field–field correlations have high values whenever the

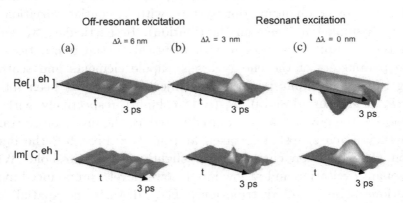

Fig. 9.5.   Longitudinal cuts of $\mathrm{Re}(I^{eh})$ (*top row*) and $\mathrm{Im}(C^{eh})$ (*bottom row*). The spectral detuning is (a) 6 nm and (b) 3 nm, (c) refers to resonant injection.

light–dipole correlations have minima and vice versa. We would like to note that, similarly, the imaginary part of the diagonal elements of the field–field correlation is strongly related to the real part of the diagonal elements of the field–dipole correlation function.

The situation is changed in the case of a small detuning (Fig. 9.5(b)). In this case, the field–field correlation shows a slow and intense response. However, the field–dipole correlation still shows fast oscillations reflecting the energetic reshuffling of the carriers by the light fields. Finally, a resonant excitation (Fig. 9.5(c)) reveals a prompt reaction of all correlation distributions to the excitation process. It is now evident that a thorough tuning of the excitation frequency may play a key role for the spatio-temporal transfer of coherence properties in optically excited active semiconductor media.

9.3 Influence of Disorder

Another key property is the degree of disorder given by material inhomogeneities (e.g. in size, energy levels and dipole matrix elements) that are typical for QD nano-structures.

Figure 9.6 shows some first results that demonstrate the influence of disorder in quantum dot structures on the spatio-temporal coherence properties snapshots of (a) the carrier density, (b) the field–field correlation function and (c) the field–dipole correlation function for a quantum dot system with a spatial variation of 5 % (assuming a Gaussian distribution). In particular, we focus on the example of a system with identical spatial fluctuations in the parameters of the energy levels, dipole elements and scattering rates. The input power of the injected pulse has been set to $0.2P_s$ (left column) and $0.7P_s$ (right column), respectively, with P_s denoting the power that is required to reduce the inversion to transparency. Each snapshot is taken at time $t = 1\,\mathrm{ps}$ after the injection. The injection current is set slightly above threshold. With resonant excitation and at low input power levels, pronounced modulations arise in all distributions. They disturb the spatial and temporal coherence and eventually lead to a reduced transfer of coherence. For high input power levels the coherence properties of

Carrier density N^{eh}

(a)

Field–field correlation I^{eh}

(b)

Field–dipole correlation C^{eh}

(c)

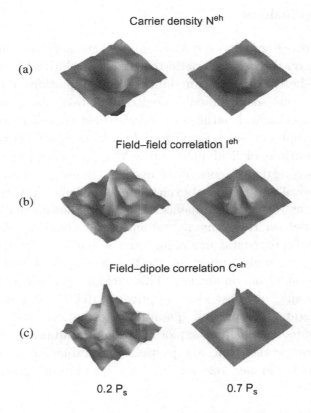

0.2 P_s 0.7 P_s

Fig. 9.6.   Influence of disorder: snapshots of the carrier density (a), field–field correlation (b) and field–dipole correlation (c) for low (*left*) and high (*right*) excitation strength (in units of the saturation power P_s).

the injected light (here: with ideal coherence properties) are via the dynamic light–matter coupling to a higher degree transferred to the medium leading to smaller modulations.

The pure material variations alone thus are only partially responsible for the coherence properties of a quantum dot nanomaterial. In addition, it is the excitation and the resulting interplay of light and matter that influences the spatio-temporal dynamics and memory properties of a given device. Disturbances induced by dot inhomogeneities may thus even be counterbalanced by excitation conditions.

9.4 Conclusions

We have investigated the spatio-temporal memory of quantum dot structures on the basis of spatio-temporal correlation functions. Our theory is based on quantum luminescence equations for spatially inhomogeneous semiconductor devices and takes into account the quantum mechanical nature of the light field as well as that of the carrier system. Our analysis of spatial coherence patterns reflected in the distributions of field–field and field–dipole correlations demonstrates a strong dependence of storage and transfer of quantum memory on optical and electrical excitation conditions. Furthermore, the distributions reveal the influence of intra dot hot-carrier relaxation and disorder on the dynamic response and coupling of light and matter. A fundamental modelling on the basis of our theory may become increasingly important for the development and design of novel functional nanomaterials. This include spatially selective signal processing, ultrafast electro-optical switching elements, spatio-spectral light trapping and quantum information processing. The shown results may help to explore parameter regimes and to design material properties that are particular suitable for an optimised memory performance relevant in future quantum communication networks.

References

Black, A.T., Thompson, J.K. and Vuletic, V., *Phys. Rev. Lett.* **95**, 133601–133604, (2005).

Chaneliere, T., Matsukevich, D.N., Jenkins, S.D., Lan, S.-Y., Kennedy, T.A.B. and Kuzmich, A., *Nature* **438**, 833–836, (2005).

Chou, C.W., Polyakov, S.V., Kuzmich, A. and Kimble, H.J., *Phys. Rev. Lett.* **92**, 213601–213604, (2004).

Eisaman, M.D., Childress, L., Andre, A., Massou, F., Zibrov, A.S. and Lukin, M.D., *Phys. Rev. Lett.* **93**, 233602–233605, (2004).

Fiurasek, J., *New J. Phys.* **8**, 192, (2006).

Gardiner, C.W. and Zoller, P., *Quantum Noise.* Springer, Berlin, (2000).

Gehrig, E. and Hess, O., *Spatio-Temporal Dynamics and Quantum Fluctuations in Semiconductor Lasers.* Springer, Heidelberg, (2003).

Julsgaard, B., Sherson, J., Cirac, J.I., Fiurasek, J. and Polzik, E.S., *Nature* **432**, 482–486, (2004).

Kuzmich, A., Bowen, W.B., Boozer, A.D., Boca, A., Chov, C.W., Duan, L.M. and Kimble, H.J., *Nature* **423**, 731–734, (2003).

Matsukevich, D.N. and Kuzmich, A., *Science* **306**, 663–666, (2004).

Santori, C., Fattal, D., Vuckovic, J., Solomon, G.S. and Yamamoto, Y., *Nature* **419**, 594–597, (2002).